Manual of Structural Kinesiology

Clem W. Thompson, Ph.D., F.A.C.S.M., who died in 1990, helped shape the field of physical education for his students and colleagues alike. Dr. Thompson wrote the fourth through the eleventh editions of *Manual of Structural Kinesiology*. He published many research papers, articles, and presentations, but he considered this book to be his most important professional accomplishment.

Dr. Thompson was a professor emeritus of physical education at Mankato State University in Mankato, Minnesota, where he served on the faculty for 25 years. He had taught at the University of Arkansas and Boston University before coming to MSU and retired from teaching in 1984. Dr. Thompson was a member of The American Alliance for Health, Physical Education, Recreation and Dance and served on various governing boards of the American Heart Association.

Dr. Thompson received his undergraduate degree from Knox College in Galesburg, Illinois, in 1938, his master's degree from the University of Illinois in 1941, and his doctorate from the University of Iowa in 1950.

Dr. Thompson was a pioneer in the campaign against smoking when it was more popular to smoke than to denounce it. He was a leader in the efforts to awaken Americans to the need to be healthy. His own outstanding personal example of physical fitness touched the lives of thousands of students and hundreds of his colleagues. His legacy of a commitment to a fit and healthy life will be perpetuated in the professional careers of them all.

Manual of Structural Kinesiology

R. T. Floyd Ed.D., ATC, CSCS

Director of Athletic Training and Sports Medicine
Professor of Physical Education and Athletic Training
Chair, Department of Physical Education and Athletic
    Training
The University of West Alabama
(formerly Livingston University)
Livingston, Alabama

Clem W. Thompson Ph.D., F.A.C.S.M.

(deceased)
Professor Physical Education, Emeritus
Mankato State University
Mankato, Minnesota

FOURTEENTH EDITION
with 187 illustrations

Boston   Burr Ridge, IL   Dubuque, IA   Madison, WI   New York   San Francisco   St. Louis
Bangkok   Bogotá   Caracas   Lisbon   London   Madrid
Mexico City   Milan   New Delhi   Seoul   Singapore   Sydney   Taipei   Toronto

McGraw-Hill Higher Education

A Division of The McGraw-Hill Companies

MANUAL OF STRUCTURAL KINESIOLOGY, FOURTEENTH EDITION

Published by McGraw-Hill, an imprint of the McGraw-Hill Companies, Inc., 1221 Avenue of the Americas, New York, NY 10020. Copyright © 2001, 1998, 1994, 1989, 1985, 1981, 1977, 1973, 1969, 1965, 1961, 1956, 1951, 1948 by The McGraw-Hill Companies, Inc. All rights reserved. No part of this publication may be reproduced or distributed in any form or by any means, or stored in a database or retrieval system, without the prior written consent of The McGraw-Hill Companies, Inc., including, but not limited to, in any network or other electronic storage or transmission, or broadcast for distance learning.

Some ancillaries, including electronic and print components, may not be available to customers outside the United States.

This book is printed on acid free paper.

3 4 5 6 7 8 9 0 QPD/QPD 0 9 8 7 6 5 4 3 2 1

ISBN 0–07–232917–3

Vice president and editor-in-chief: *Kevin T. Kane*
Executive editor: *Vicki Malinee*
Senior developmental editor: *Melissa Martin*
Senior marketing manager: *Pamela S. Cooper*
Production manager: *Mary E. Powers*
Production supervisor: *Enboge Chong*
Coordinator of freelance design: *Rick D. Noel*
Cover designer: *Crispin Prebys*
Cover image: © *Tony Stone, image no. BB1165–005 "Man Clearing Hurdle" by Alan Thornton*
Senior photo research coordinator: *Carrie K. Burger*
Supplement coordinator: *Brenda A. Ernzen*
Compositor: *Interactive Composition Corp.*
Typeface: *10.5/12.5 Garamond*
Printer: *Quebecor Printing Book Group/Dubuque, IA*

The credits section for this book begins on page 272 and is considered an extension of the copyright page.

The author and publisher disclaim any responsibility for any adverse effects or consequences from the misapplication or injudicious use of information contained within this text.

Library of Congress Cataloging-in-Publication Data

Floyd, R. T.
    Manual of structural kinesiology / R. T. Floyd, Clem W. Thompson. —14th ed.
        p.   ; cm.
    Includes bibliographical references and index.
    ISBN 0–07–232917–3 (alk. paper)
    1. Kinesiology. 2. Human locomotion. 3. Muscles. I. Thompson, Clem W. II. Title.
[DNLM: 1. Movement—physiology. 2. Kinesiology, Applied. 3. Muscles—physiology.
WE 103 F645m 2001]
QP303 .T58   2001   00–32864
612.7′6—dc21   CIP

www.mhhe.com

To

my family,
Lisa, Robert Thomas, Jeanna, Rebecca, and Kate,
who understand, support, and allow me to
pursue my profession

R.T.F.

Preface

In revising this edition, I have attempted to further refine the chapters for consistency and to improve the overall completeness and accuracy of the text. As with previous revisions, I have attempted to maintain the successful presentation approach the late Dr. Clem Thompson established from 1961 through 1989. I first used this book as an undergraduate and later in my teaching over the years. Having developed great respect for this text and Dr. Thompson's style, I intend to continue to preserve the effectiveness of this time-honored text, while adding material pertinent to the professions working with today's ever growing physically active population. I hope I have maintained a clear, concise, and simplistic presentation method supplemented with applicable information gained through my career experiences.

This text, now in its 53rd year, has undergone many revisions over the years. My goal continues to make the material as applicable as possible to everyday physical activity and to make it more understandable and easier to use for the student. I have tried to improve the consistency of the illustrations, while adding more directional arrows for joint movements and student exercises. I challenge kinesiology students and professionals while reading to immediately apply the content of this text to physical activities with which they are familiar. I hope that the student will simultaneously palpate his or her own moving joints and contracting muscles while reading to gain application. Simultaneously, I encourage students to palpate the joints and muscles of fellow students to gain a better appreciation of the wide range of normal anatomy and, when possible, appreciate the variation from normal found in injured and pathological musculoskeletal anatomy.

Audience

Applied kinesiologists, athletic trainers, athletic coaches, physical educators, physical therapists, health club instructors, strength and conditioning specialists, personal trainers, massage therapists, physicians, and others who are responsible for improving and maintaining the muscular strength, endurance, flexibility, and overall health of individuals will benefit from this text.

With the tremendous growth in the number of participants in an ever increasing spectrum of physical activity, it is imperative that medical, health, and education professionals involved in providing instruction and information to the physically active be correct and accountable for the teachings that they provide. The variety of exercise machines, techniques, strengthening and flexibility programs, and training programs is constantly expanding, but the musculoskeletal system is constant in its design and architecture. Regardless of the goals sought or the approaches used in exercise activity, the human body is the basic ingredient and must be thoroughly understood and considered to maximize performance capabilities and minimize undesirable results. Most advances in exercise science continue to result from a better understanding of the body and how it functions. I believe that an individual in this field can never learn enough about structure and function of the human body.

Those who are charged with the responsibility of providing instruction to the physically active will find this text a helpful and valuable resource in their never-ending quest for knowledge and understanding of human movement.

New to this edition

Major changes to this edition include the revision of several illustrations, as well as some expansion of material presented. I have changed the individual muscle illustrations for clarity by placing the directional arrows for joint movement on the illustration containing the muscle, while moving the origin and insertion labeling to the illustration containing the location of these attachments. Some other illustrations have been replaced to improve understanding. Particular emphasis has been placed on correcting the labeling errors that occurred in the production of the previous edition. The movement analysis examples in Chapters 6 and 11 have been carefully reviewed to ensure accuracy, with added emphasis on the grouping of muscles by their actions in functional movement of the joints.

In the eight chapters that address specific body parts, exercises have been added for students to use in exploring the *Dynamic Human* CD-ROM, which accompanies this text. In the captions of the muscle illustrations, the view that is portrayed has been added throughout for orientation purposes. A few student objectives have been added or revised when appropriate to reflect changes, and several new references have been added to reflect more current resources. Chapter 1, "Foundations of Structural Kinesiology," has been expanded significantly to include more information regarding neuromuscular function and the role of muscles in joint movement. Additionally, some revision of the material on planes and axes has occurred, along with the inclusion of the planes in which movement occurs in the joint movement sections. Some additional synonyms for these movements have been added. An appendix has been added to provide a summary of the diarthrodial joint types and their ranges of motion. Student exercises for the *Dynamic Human* CD-ROM have been added here as well.

As mentioned previously, numerous illustrations have been revised for clarity. Several of these are located in Chapter 4, "The Elbow and Radioulnar Joints;" Chapter 5, "The Wrist and Hand Joints;" Chapter 7, "The Hip joint and Pelvic Girdle;" and Chapter 10, "The Trunk and Spinal Column."

A few biomechanical concepts have been added, along with some revision of the existing ones, in Chapter 12, "Basic Biomechanical Factors and Concepts." The glossary has been expanded to include the additional terminology new to this edition. Finally, web sites are included at the end of each chapter.

Acknowledgments

I am very appreciative of the comments, ideas, and suggestions provided by the six reviewers which have served as an extremely helpful guide in this revision:

Gail Arnold
University of Massachusetts–Boston

Sandra Cole
Stephen F. Austin State University

Tibor Hortobagyi
East Carolina University

Marianne McAdam
Eastern Kentucky University

Virginia Overdorf
William Paterson University

Carole Zebas
University of Kansas

I would also like to thank Brad Montgomery, MAT, ATC; Kurt Behrhorst, MAT, ATC; Alex Dibbley, ATC; Stephen Guthrie, ATC; and Jake Jordan, ATC, of the University of West Alabama for their advice and input throughout this revision. Their assistance and suggestions have helped significantly. I also acknowledge John Hood and Lisa Floyd of Birmingham and Livingston, Alabama, respectively, for the fine photographs. Special thanks to Linda Kimbrough of Birmingham, Alabama, for her superb illustrations and insight. I appreciate the models for the photographs, Marcus Shapiro, Zina Pruitt, and Darrell Locket. My thanks also go to Melissa Martin, Vicki Malinee, and Mary Powers of the McGraw-Hill staff, who have been most helpful in their assistance and suggestions in preparing the manuscript for publication.

R.T. Floyd

Contents

Foundations of structural kinesiology

Objectives

- To review the anatomy of the skeletal and muscle systems

- To review and understand the terminology used to describe body part locations, reference positions, and anatomical directions

- To review the planes of motion and their respective axes of rotation in relation to human movement

- To describe and understand the various types of bones and joints in the human body and their characteristics

- To describe and demonstrate the joint movements

- To learn and understand the different types of muscle contractions and how muscles function in joint movement

- To learn and understand basic neuromuscular concepts in relation to the ways in which muscles function in joint movement

Structural kinesiology is the study of muscles as they are involved in the science of movement. Both skeletal and muscular structures are involved. Bones are different sizes and shapes—particularly at the joints, which allow or limit movement. Muscles vary greatly in size, shape, and structure from one part of the body to another.

More than 600 muscles are found in the human body. In this book, an emphasis is placed on the larger muscles that are primarily involved in movement of the joints. Details related to many of the small muscles located in the hands, feet, and spinal column are provided to a lesser degree.

However, anatomists, coaches, strength and conditioning specialists, personal trainers, nurses, physical educators, physical therapists, physicians, athletic trainers, massage therapists, and others in health-related fields should have an adequate knowledge and understanding of all the large muscle groups, so they can teach others how to strengthen, improve, and maintain these parts of the human body. This knowledge forms the basis of the exercise programs that should be followed to strengthen and maintain all of the muscles. In most cases, exercises that involve the larger primary movers also involve the smaller muscles.

Fewer than 100 of the largest and most important muscles, primary movers, are considered in this text. Some small muscles in the human body, such as the multifidus, plantaris, scalenus, and serratus posterior, are omitted, since they are exercised with other, larger primary movers. In addition, most small muscles of the hands and feet are not given the full attention provided to the larger muscles. Many small muscles of the spinal column are not considered in full detail.

Kinesiology students frequently become so engrossed in learning individual muscles that they lose sight of the total muscular system. They miss the "big picture"—that muscle groups move joints in given movements necessary for bodily movement and skilled performance. Although it is vital to learn the small details of muscle attachments, it is even more critical to be able to apply the information to real-life situations. Once the information can be applied in a useful manner, the specific details are usually much easier to understand and appreciate.

Skeletal and muscular systems

Fig. 1.1 shows anterior and posterior views of the skeletal system. Two hundred six bones make up the skeletal system, which provides support and protection for other systems of the body and provides for attachments of the muscles to the bones by which movement is produced. Additional skeletal functions are mineral storage and hemopoiesis, which involves blood cell formation in the red bone marrow. The skeleton may be divided into the appendicular and the axial skeleton. The appendicular skeleton is composed of the appendages, or the upper and lower extremities, and the shoulder and pelvic girdles, while the axial skeleton consists of the skull, vertebral col-

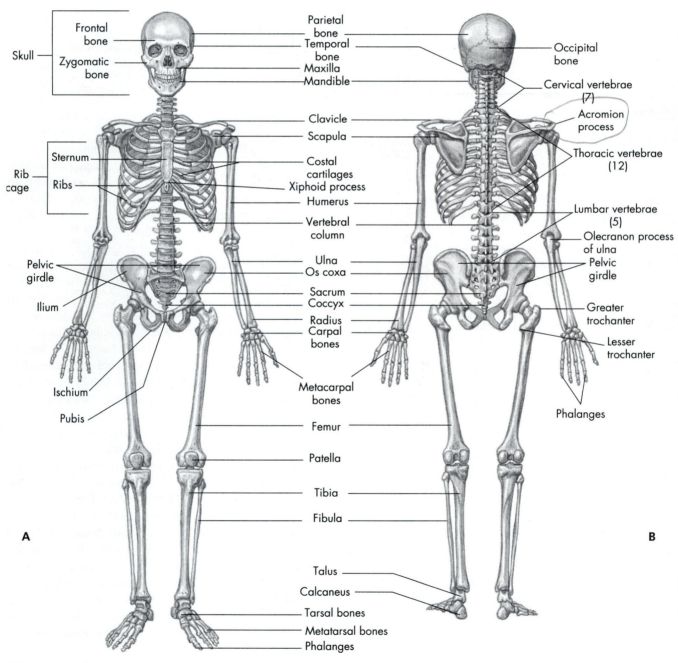

FIG. 1.1 ● Skeleton. **A**, Anterior view; **B**, posterior view.

From Van De Graff KM: *Human Anatomy*, ed 4, 1995, McGraw-Hill Companies, Inc., New York.

umn, ribs, and sternum. Most students who take this course will have had a course in human anatomy, but a brief review is desirable before beginning the study of kinesiology. Other chapters provide additional information and more detailed illustrations of specific bones.

The total superficial muscular system is shown in Figs. 1.2 and 1.3. Any figure is limited, since many muscles are deep to the surface muscles. Still, these figures will help provide a better overview of the entire superficial muscular structure.

Muscles shown in these figures, and many other muscles, will be studied in more detail as each joint of the body is considered in other chapters of this book.

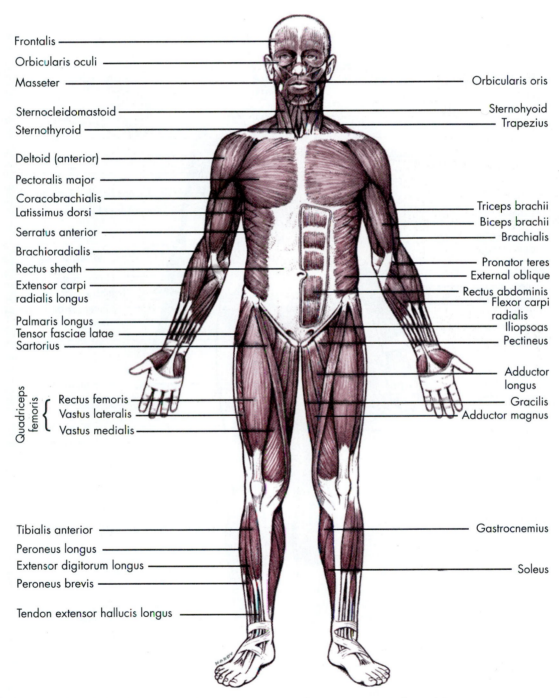

FIG. 1.2 ● Muscles of the human body, anterior view from anatomical position.

From Thibodeau GA: *Anatomy and physiology*, St. Louis, 1987, Mosby.

Reference positions

It is crucial for kinesiology students to begin with a reference point in order to better understand the musculoskeletal system, its planes of motion, joint classification, and joint movement terminology. Two reference positions may be used as a basis from which to describe joint movements. The *anatomical position* is the most widely used and accurate for all aspects of the body. Fig. 1.2 demonstrates this reference position, with the subject standing in an upright posture, facing straight ahead, feet parallel and close, and the palms facing forward. The *fundamental position* is essentially the same as the anatomical position, except that the arms are at the sides and facing the body.

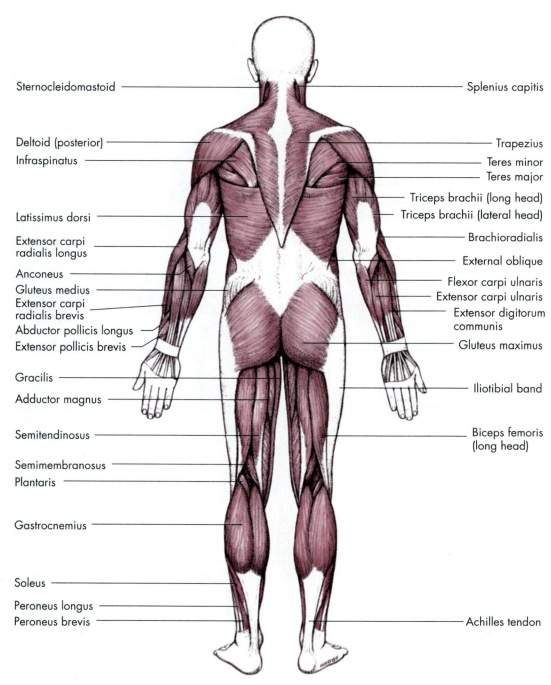

Sternocleidomastoid

Deltoid (posterior)

Infraspinatus

Latissimus dorsi

Extensor carpi radialis longus

Anconeus

Gluteus medius

Extensor carpi radialis brevis

Abductor pollicis longus

Extensor pollicis brevis

Gracilis

Adductor magnus

Semitendinosus

Semimembranosus

Plantaris

Gastrocnemius

Soleus

Peroneus longus

Peroneus brevis

Splenius capitis

Trapezius

Teres minor

Teres major

Triceps brachii (long head)

Triceps brachii (lateral head)

Brachioradialis

External oblique

Flexor carpi ulnaris

Extensor carpi ulnaris

Extensor digitorum communis

Gluteus maximus

Iliotibial band

Biceps femoris (long head)

Achilles tendon

FIG. 1.3 ● Muscles of the human body, posterior view.

From Thibodeau GA: *Anatomy and physiology*, St. Louis, 1987, Mosby.

Anatomical directional terminology

FIG. 1.4

Anterior
in front or in the front part

Anteroinferior
in front and below

Anterolateral
in front and to the side, especially the outside

Anteromedial
in front and toward the inner side or midline

Anteroposterior
relating to both front and rear

Anterosuperior
in front and above

Caudal
below in relation to another structure; inferior

Cephalic
above in relation to another structure; higher, superior

Contralateral
pertaining or relating to the opposite side

Deep
beneath or below the surface; used to describe relative depth or location of muscles or tissue

Distal
situated away from the center or midline of the body, or away from the point of origin

Dorsal
relating to the back; posterior

Inferior
(infra) below in relation to another structure; caudal

Ipsilateral
on the same side

Lateral
on or to the side; outside, farther from the median or midsagittal plane

Medial
relating to the middle or center; nearer to the medial or midsagittal plane

Posterior
behind, in back, or in the rear

Posteroinferior
behind and below; in back and below

Posterolateral
behind and to one side, specifically to the outside

Posteromedial
behind and to the inner side

Posterosuperior
behind and at the upper part

Prone
the body lying face downward; stomach lying

Proximal
nearest the trunk or the point of origin

Superficial
near the surface; used to describe relative depth or location of muscles or tissue

Superior
(supra) above in relation to another structure; higher, cephalic

Supine
lying on the back; face upward position of the body

Ventral
relating to the belly or abdomen

Volar
relating to palm of the hand or sole of the foot

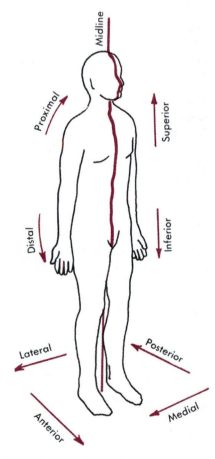

FIG. 1.4 ● Anatomical directions.

From Arnheim DD, Prentice WE: *Principles of athletic training*, ed 9, St. Louis, 1997, WCB/McGraw-Hill.

Planes of motion

When studying the various joints of the body and analyzing their movements, it is helpful to characterize them according to specific planes of motion (Fig. 1.5). A plane of motion may be defined as an imaginary two-dimensional surface through which a limb or body segment is moved.

There are three specific planes of motion in which the various joint movements can be classified. Although each specific joint movement can be classified as being in one of the three planes of motion, our movements are usually not totally in one specific plane but occur as a combination of motions from more than one plane. These movements from the combined planes may be described as occurring in diagonal or oblique planes of motion.

Sagittal, anteroposterior, or AP plane

The sagittal, anteroposterior, or AP plane bisects the body from front to back, dividing it into right and left symmetrical halves. Generally, flexion and extension movements such as biceps curls, knee extensions, and sit-ups occur in this plane.

Frontal, lateral, or coronal plane

The frontal plane, also known as the lateral or coronal plane, bisects the body laterally from side to side, dividing it into front and back halves. Abduction and adduction movements such as jumping jacks and spinal lateral flexion occur in this plane.

Transverse or horizontal plane

The transverse plane divides the body horizontally into superior and inferior halves. Generally, rotational movements such as pronation, supination, and spinal rotation occur in this plane.

Diagonal or oblique plane

The diagonal or oblique plane is a combination of more than one plane. In reality, most of our movements in sporting activities fall somewhere less than parallel or perpendicular to the previously described planes and occur in a diagonal plane.

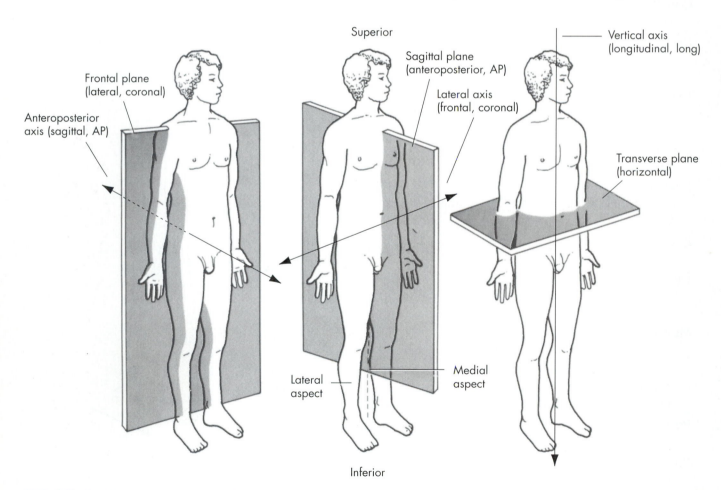

FIG. 1.5 • Planes of motion and axes of rotation.

Modified from Booher JM, Thibodeau GA: *Athletic injury assessment*, ed 3, St. Louis, 1994, Mosby.

Axes of rotation

As movement occurs in a given plane, the joint moves or turns about an axis that has a 90-degree relationship to that plane. The axes are named in relation to their orientation (Fig. 1.5). Table 1.1 lists the planes of motions with their axes of rotation.

Lateral, frontal, or coronal axis

The lateral, frontal, or coronal axis has the same directional orientation as the frontal plane of motion and runs from side to side at a right angle to the sagittal plane of motion.

Anteroposterior or sagittal axis

The anteroposterior axis has the same directional orientation as the sagittal plane of motion and runs from front to back at a right angle to the frontal plane of motion.

Vertical or longitudinal axis

The vertical axis, also known as the longitudinal or long axis, runs straight down through the top of the head and is at a right angle to the transverse plane of motion.

TABLE 1.1 • **Planes of motion and their axes of rotation**

Plane	Axis
Sagittal (anteroposterior or AP)	Lateral (frontal or coronal)
Frontal (lateral or coronal)	Anteroposterior (sagittal or AP)
Transverse (horizontal)	Vertical (longitudinal or long)

Types of bones

Bones vary greatly in shape and size but can be categorized into five major categories (Fig. 1.6).

Long bones *diaphysis*

These bones are composed of a long cylindrical shaft with relatively wide, protruding ends. The shaft contains the medullary canal. Examples include phalanges, metatarsals, metacarpals, tibia, fibula, femur, radius, ulna, and humerus.

Short bones

Small, cubical shaped, solid bones, which usually have a proportionally large articular surface in order to articulate with more than one bone. Short bones provide some shock absorption and include the carpals and tarsals.

Flat bones

Usually with a curved surface and varying from thick, where tendons attach, to very thin. Flat bones generally provide protection and include the ilium, ribs, sternum, clavicle, and scapula.

Irregular bones

Irregularly shaped bones serve a variety of purposes and include the bones throughout the entire spine and the ischium, pubis, and maxilla.

Sesamoid bones

These are small bones embedded within the tendon of a musculotendinous unit that provide protection as well as improve the mechanical advantage of musculotendinous units. In addition to the patella, there are small sesamoid bones within the flexor tendons of the great toe and the thumb.

epiphysis is formed from cancellous, spongy & trabecular bone.
Fibrous membrane cover inside surface of the cortex = endosteum

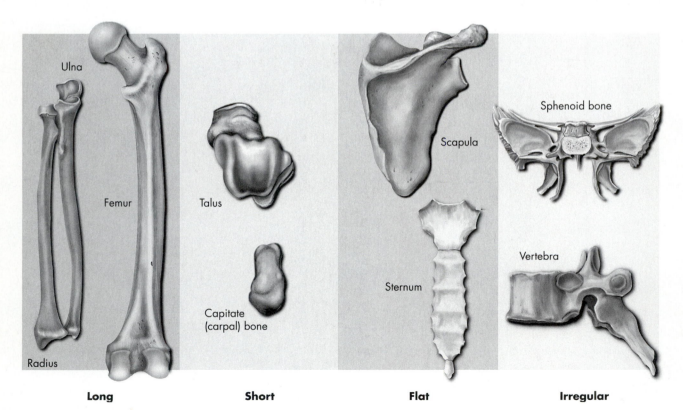

Long   **Short**   **Flat**   **Irregular**

Ulna · Femur · Radius · Talus · Capitate (carpal) bone · Scapula · Sternum · Sphenoid bone · Vertebra

FIG. 1.6 ● Classification of bones by shape.

Types of joints

The articulation of two or more bones allows various types of movement. The extent and type of movement determine the name applied to the joint. Bone structure limits the kind and amount of movement in each joint. Some joints have no movement, others are only very slightly movable, and others are freely movable with a variety of movement ranges. The type and range of movements are similar in all humans, but the freedom, range, and vigor of movements vary between individuals, due to limitations imposed by ligaments, muscles, and slight variations in joint structure.

The articulations are grouped into three classes based on their structure and the amount of movement possible.

Synarthrodial (immovable) joints (Fig. 1.7)

Found in the sutures of the cranial bones and sockets of the teeth

Amphiarthrodial (slightly movable) joints (Fig. 1.8)

Structurally, these articulations are divided into two groups:

Syndesmosis

Type of joint held together by strong ligamentous structures that allow minimal movement between the bones. Examples are the coracoclavicular joint and the inferior tibiofibular joint.

Synchondrosis

Type of joint separated by a fibrocartilage that allows very slight movement between the bones. Examples are the symphysis pubis and the costochondral joints of the ribs with the sternum.

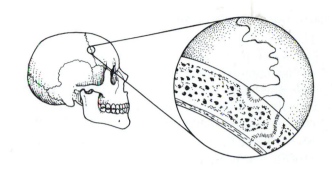

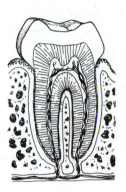

FIG. 1.7 ● Synarthrodial joints.

Modified from Booher JM, Thibodeau GA: *Athletic injury assessment,* ed 3, St. Louis, 1994, Mosby.

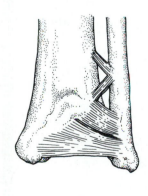

A

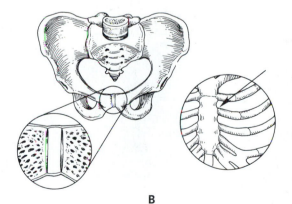

B

FIG. 1.8 ● Amphiarthrodial joints. **A,** Syndesmosis joint; **B,** synchondrosis joint.

Modified from Booher JM, Thibodeau GA: *Athletic injury assessment,* ed 3, St. Louis, 1994, Mosby.

Diarthrodial (freely movable) joints (Fig. 1.9)

Di-arthrodial joints, also known as synovial joints, are freely movable. A sleevelike covering of ligamentous tissue, known as the *joint capsule*, surrounds the bony ends forming the joints. This ligamentous capsule is lined with a thin vascular synovial capsule that secretes synovial fluid to lubricate the area inside the joint capsule, known as the *joint cavity*. In certain areas, the capsule is thickened to form tough, nonelastic ligaments that provide additional support against abnormal movement or joint opening. These ligaments vary in location, size, and strength, depending on the particular joint.

In many cases, additional ligaments, not continuous with the joint capsule, provide further support. The articular surfaces on the ends of the bones inside the joint cavity are covered with layers of articular, or *hyaline cartilage*. This resilient cartilage absorbs shock to protect the bone it covers. When the joint surfaces are unloaded or distracted, this articular cartilage slowly absorbs a slight amount of the joint synovial fluid, only to slowly secrete it during subsequent weight bearing and compression. Additionally, some diarthrodial joints have a fibrocartilage disk between their articular surfaces. Structurally, this type of articulation can be divided into six groups, as shown in Fig. 1.10.

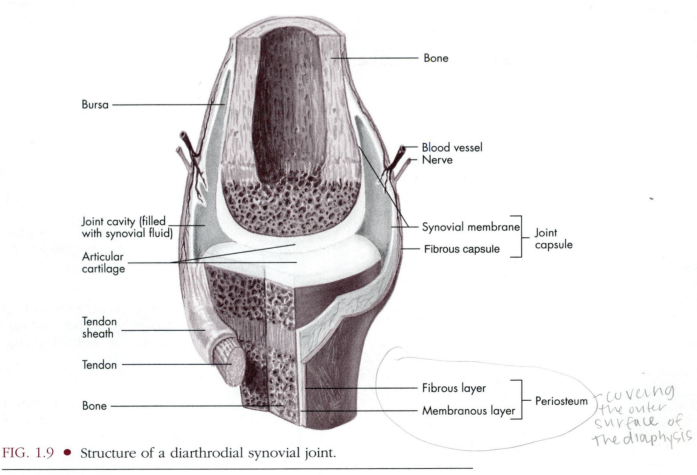

FIG. 1.9 ● Structure of a diarthrodial synovial joint.

From Seeley RR, Stephens TD, Tate P: *Anatomy & physiology*, ed 3, St. Louis, 1995, Mosby-Year Book.

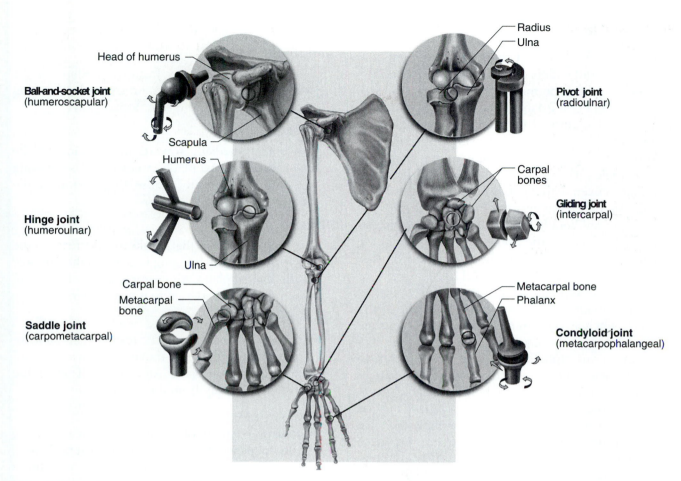

Ball-and-socket joint
(humeroscapular)

Head of humerus

Scapula

Humerus

Hinge joint
(humeroulnar)

Ulna

Carpal bone

Metacarpal bone

Saddle joint
(carpometacarpal)

Radius

Ulna

Pivot joint
(radioulnar)

Carpal bones

Gliding joint
(intercarpal)

Metacarpal bone

Phalanx

Condyloid joint
(metacarpophalangeal)

FIG. 1.10 ● Types of diarthrotic or synovial joints.

Arthrodial (gliding) joint

Characterized by two flat, bony surfaces that butt against each other. This type of joint permits limited gliding movement. Examples are the carpal bones of the wrist and the tarsometatarsal joints of the foot.

Condyloidal (biaxial ball-and-socket) joint

This is a type of joint in which the bones permit movement in two planes without rotation. Examples are the wrist between the radius and the proximal row of the carpal bones or the second, third, fourth, and fifth metacarpophalangeal joints.

Enarthrodial (multiaxial ball-and-socket) joint

This is a type of joint that permits movement in all planes. Examples are the shoulder (glenohumeral) and hip joints.

Ginglymus (hinge) joint

This is a type of joint that permits a wide range of movement in only one plane. Examples are the elbow, ankle, and knee joints.

Sellar (saddle) joint

This type of reciprocal reception is found only in the thumb at the carpometacarpal joint and permits ball-and-socket movement, with the exception of rotation.

Trochoidal (pivot) joint

This is a type of joint with a rotational movement around a long axis. An example is the rotation of the radius at the radioulnar joint.

Movements in joints

In many joints, several different movements are possible. Some joints permit only flexion and extension; others permit a wide range of movements, depending largely on the joint structure. The specific amount of movement possible in a joint may be measured by using an instrument known as a *goniometer* to compare the change in joint angles. The goniometer axis, or hinge point, is placed even with the axis of rotation at the joint line. As the joint is moved, both arms of the goniometer are held in place either along or parallel to the long axis of the bones on either side of the joint. The joint angle can then be read from the goniometer, as shown in Fig. 1.11.

Some movement terms may be used to describe motion at several joints throughout the body, whereas other terms are relatively specific to a joint or group of joints (Fig. 1.12). Rather than list the terms alphabetically, we have chosen to group them according to the body area and pair them with opposite terms where applicable. Additionally, prefixes such as *hyper-* and *hypo-* may be combined with these terms to emphasize excessive or reduced motion, respectively. Of these combined terms, *hyperextension* is the most commonly used.

General

Abduction

Lateral movement away from the midline of the trunk in the frontal plane. An example is raising the arms or legs to the side horizontally.

Adduction

Movement medially toward the midline of the trunk in the frontal plane. An example is lowering the arm to the side or the thigh back to the anatomical position.

Flexion

Bending movement that results in a decrease of the angle in a joint by bringing bones together, usually in the sagittal plane. An example is the elbow joint when the hand is drawn to the shoulder.

Extension

Straightening movement that results in an increase of the angle in a joint by moving bones apart, usually in the sagittal plane. An exam-ple is when the hand moves away from the shoulder.

Circumduction

Circular movement of a limb that delineates an arc or describes a cone. It is a combination of flexion, extension, abduction, and adduction. Sometimes referred to as circumflexion. An example is when the shoulder joint and the hip joint move in a circular fashion around a fixed point.

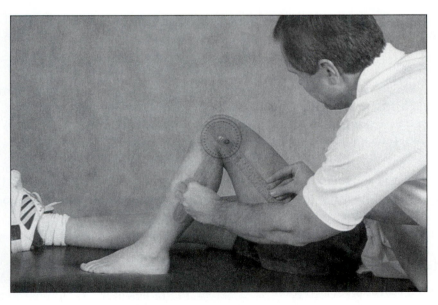

FIG. 1.11 ● Goniometric measurement of knee joint flexion.

From Arnheim DD, Prentice WE: *Principles of athletic training*, ed 9, St. Louis, 1997, WCB/McGraw-Hill.

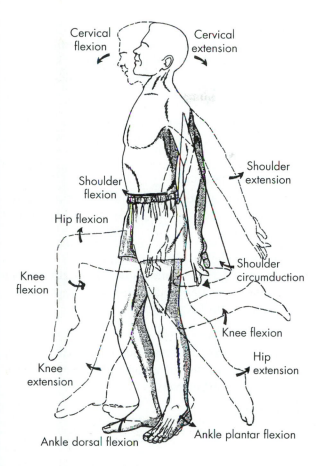

FIG. 1.12 • Joint movements.

Diagonal abduction

Movement by a limb through a diagonal plane away from the midline of the body.

Diagonal adduction

Movement by a limb through a diagonal plane toward and across the midline of the body.

External rotation

Rotary movement around the longitudinal axis of a bone away from the midline of the body. Occurs in the transverse plane and is also known as rotation laterally, outward rotation, and lateral rotation.

Internal rotation

Rotary movement around the longitudinal axis of a bone toward the midline of the body. Occurs in the transverse plane and is also known as rotation medially, inward rotation, and medial rotation.

Ankle and foot

Eversion

Turning the sole of the foot outward or laterally in the frontal plane; abduction. An example is

standing with the weight on the inner edge of the foot.

Inversion

Turning the sole of the foot inward or medially in the frontal plane; adduction. An example is standing with the weight on the outer edge of the foot.

Dorsal flexion (dorsiflexion)

Flexion movement of the ankle that results in the top of the foot moving toward the anterior tibia bone in the sagittal plane.

Plantar flexion

Extension movement of the ankle that results in the foot and/or toes moving away from the body in the sagittal plane.

Radioulnar joint

Pronation

Internally rotating the radius in the transverse plane so that it lies diagonally across the ulna, resulting in the palm-down position of the forearm.

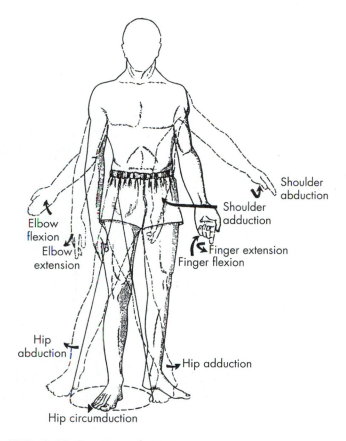

FIG. 1.12 Continued • Joint movements.

Supination

Externally rotating the radius in the transverse plane so that it lies parallel to the ulna, resulting in the palm-up position of the forearm.

Shoulder girdle and shoulder joint

Depression

Inferior movement of the shoulder girdle in the frontal plane. An example is returning to the normal position from a shoulder shrug.

Elevation

Superior movement of the shoulder girdle in the frontal plane. An example is shrugging the shoulders.

Horizontal abduction

Movement of the humerus in the horizontal plane away from the midline of the body. Also known as horizontal extension or transverse abduction.

Horizontal adduction

Movement of the humerus in the horizontal plane toward the midline of the body. Also known as horizontal flexion or transverse adduction.

Protraction (abduction)

Forward movement of the shoulder girdle in the horizontal plane away from the spine. Abduction of the scapula.

Retraction (adduction)

Backward movement of the shoulder girdle in the horizontal plane toward the spine. Adduction of the scapula.

Rotation downward

Rotary movement of the scapula in the frontal plane with the inferior angle of the scapula moving medially and downward.

Rotation upward

Rotary movement of the scapula in the frontal plane with the inferior angle of the scapula moving laterally and upward.

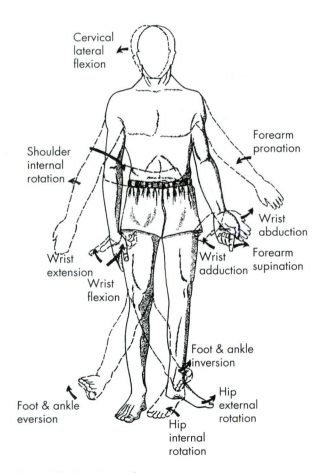

FIG. 1.12 Continued ● Joint movements.

Spine

Lateral flexion (side bending)

Movement of the head and/or trunk in the frontal plane laterally away from the midline. Abduction of the spine.

Reduction

Return of the spinal column in the frontal plane to the anatomic position from lateral flexion. Adduction of the spine.

Wrist and hand

Palmar flexion

Flexion movement of the wrist in the sagittal plane with the volar or anterior side of the hand moving toward the anterior side of the forearm.

Radial flexion (radial deviation)

Abduction movement at the wrist in the frontal plane of the thumb side of the hand toward the forearm.

Ulnar flexion (ulnar deviation)

Adduction movement at the wrist in the frontal plane of the little finger side of the hand toward the forearm.

Opposition of the thumb

Diagonal movement of the thumb across the palmar surface of the hand to make contact with the fingers.

These movements are considered in detail in the chapters that follow as they apply to the individual joints.

Combinations of movements can occur. Flexion or extension can occur with abduction, adduction, or rotation.

Muscle terminology

Muscle contraction produces the force that causes joint movement in the human body. Locating the muscles, their proximal and distal attachments, and their relationship to the joints they cross is critical to determining the effects that muscles have on the joints. It is also necessary to understand certain terms as body movement is considered.

Intrinsic

Pertaining usually to muscles within or belonging solely to the body part on which they act. The small intrinsic muscles found entirely within the hand are examples.

Extrinsic

Pertaining usually to muscles that arise or originate outside of (proximal to) the body part on which they act. The forearm muscles that attach proximally on the distal humerus and insert on the fingers are examples.

When a particular muscle contracts, it tends to pull both ends toward the middle, or belly, of the muscle. Consequently, if neither of the bones to which a muscle is attached were stabilized, then both bones would move toward each other upon contraction. The more common case, however, is that one bone is more stabilized by a variety of factors and as a result the less stabilized bone usually moves toward the more stabilized bone upon contraction.

Origin

Proximal attachment, generally considered the least movable part or the part that attaches closest to the midline or center of the body.

Insertion

Distal attachment, generally considered the most movable part or the part that attaches farthest from the midline or center of the body.

As an example, the biceps brachii muscle in the arm has its origin (least movable bone) on the scapula and its insertion (most movable bone) on the radius. In some movements, this process can be reversed. An example of this reversal can be seen in the pull-up, where the radius is relatively stable and the scapula moves up. The biceps brachii would be an extrinsic muscle of the elbow, whereas the brachialis would be intrinsic to the elbow. For each muscle studied, the origin and insertion are indicated.

Types of muscle contraction

All muscle contractions can be classified as being either isometric or isotonic. An isometric contraction occurs when tension is developed within the muscle but the joint angles remain constant. Isometric contractions may be thought of as static contractions because a significant amount of tension may be developed in the muscle to maintain the joint angle in a relatively static or stable position.

Isotonic contractions involve the muscle developing tension to either cause or control joint movement and may be thought of as dynamic contractions because the varying degrees of tension in the muscles are causing the joint angles to change. The isotonic type of muscle contraction is classified further as being either concentric or eccentric on the basis of whether shortening or lengthening occurs. Fig. 1.13, *A* illustrates isotonic and Fig. 1.13, *B* demonstrates isometric contractions.

It is also important to note that movement may occur at any given joint without any muscle contraction whatsoever. This movement is referred to as passive and is solely due to external forces such as those applied by another person, an object, resistance, or the force of gravity in the presence of muscle relaxation.

Concentric contractions involve the muscle developing tension as it shortens, whereas eccentric contractions involve the muscle lengthening under tension. Actually, the term *contraction* is somewhat contradictory in regard to eccentric muscle activity, since the muscle is really lengthening while maintaining considerable tension. Eccentric muscle action is perhaps more correct.

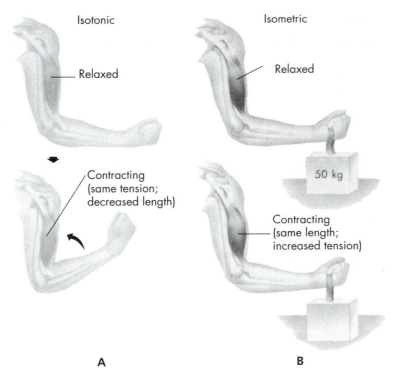

FIG. 1.13 ● Isotonic and isometric contraction. **A,** In an isotonic contraction, the muscle shortens, producing movement. **B,** In an isometric contraction, the muscle pulls forcefully against a load but does not shorten.

From Booher JM, Thibodeau GA: *Athletic injury assessment,* ed 3, St. Louis, 1994, Mosby.

Isometric contraction

Occurs when tension is developed within the muscle but no appreciable change occurs in the joint angle or in the length of the muscle; also known as a static contraction. The force developed by the muscle is equal to that of the resistance.

Isotonic contraction

Occurs when tension is developed in the muscle while it either shortens or lengthens; also known as a dynamic contraction and may be classified as being either concentric or eccentric. The force developed by the muscle is either greater or less than that of the resistance.

Concentric contraction

Involves the muscle developing tension as it shortens and occurs when the muscle develops enough force to overcome the applied resistance. Concentric contractions may be thought of as causing movement against gravity or resistance and are described as being positive contractions. The force developed by the muscle is greater than that of the resistance, results in the joint angle changing in the direction of the applied muscle force, and causes the body part to move against gravity or external forces.

Eccentric contraction (muscle action)

Involves the muscle lengthening under tension and occurs when the muscle gradually lessens in tension to control the descent of the resistance. The weight or resistance may be thought of as overcoming the muscle contraction but not to the point that the muscle cannot control the descending movement. Eccentric muscle actions control movement with gravity or resistance and are described as negative contractions. The force developed by the muscle is less than that of the resistance, results in the joint angle changing in the direction of the resistance or external force, and causes the body part to move with gravity or external forces (resistance). Some authorities refer to this as a muscle action instead of a contraction, since the muscle is lengthening as opposed to shortening.

Table 1.2 explains the various types of contraction and resultant joint movements. The terminology used in defining and describing these actions is included.

Various exercises may use any one or all of the contraction types for muscle development. Development of exercise machines has resulted in another type of muscle exercise known as *isokinetics*. Isokinetics is not another type of contraction, as some authorities have described; rather, it is a specific technique that may use any or all of the different types of contractions. Isokinetics is a type of dynamic exercise usually using concentric and/or eccentric muscle contractions in which the speed (or velocity) of movement is constant and muscular contraction (ideally, maximum contraction) occurs throughout the movement. Biodex, Cybex, Lido, and other types of apparatus are engineered to allow this type of exercise.

Students well educated in kinesiology should be qualified to prescribe exercises and activities for the development of large muscles and muscle groups in the human body. They should be able to read the description of an exercise or observe an exercise and immediately know the most important muscles being used. Descriptive terms of how muscles function in joint movements follow.

Role of muscles

Agonist

Generally described as muscles that, when contracting concentrically, cause joint motion through a specified plane of motion; known as primary or prime movers, or muscles most involved.

Antagonist

Muscles that are usually located on the opposite side of the joint from the agonist and have the opposite action; known as contralateral muscles, they work in cooperation with agonist muscles by relaxing and allowing movement, but, when contracting concentrically, perform the opposite joint motion of the agonist.

TABLE 1.2 • **Muscle contraction and movement matrix**

Definitive and descriptive factors	Types of contraction			Movement without contraction
	Isometric	Isotonic		
		Concentric	Eccentric	
Agonist muscle length	No appreciable change	Shortening ➡⬅	Lengthening ⬅➡	Dictated solely by gravity and/or external forces
Antagonist muscle length	No appreciable change	Lengthening ⬅➡	Shortening ➡⬅	Dictated solely by gravity and/or external forces
Joint angle changes	No appreciable change	In direction of force	In direction of external force (resistance)	Dictated solely by gravity and/or external forces
Direction of body part	Against immovable object or matched external force (resistance)	Against gravity and/or other external force (resistance)	With gravity and/or other external force (resistance)	Consistent with gravity and/or other external forces
Motion	Pressure (force) applied, but no resulting motion	Causes motion	Controls motion	Either no motion or passive motion occurs as a result of gravity and/or other external forces
Description	Static Fixating	Dynamic shortening Positive work	Dynamic lengthening Negative work	Passive Relaxation
Applied muscle force versus resistance	Force = resistance	Force > resistance	Force < resistance	No force, all resistance
Speed relative to gravity or applied resistance including inertial forces	Equal to speed of applied resistance	Faster than the inertia of the resistance	Slower than the speed of gravity or applied inertial forces	Consistent with inertia of applied external forces or the speed of gravity
Acceleration/ deceleration	Zero acceleration	Acceleration ↗	Deceleration ↘	Either zero or acceleration consistent with applied external forces
Descriptive symbol	(=)	(+)	(−)	(0)

Stabilizers

Muscles that surround the joint or body part and contract to fixate or stabilize the area to enable another limb or body segment to exert force and move; known as fixators, they are essential in establishing a relatively firm base for the more distal joints to work from when carrying out movements.

Synergist

Muscles that assist in the action of the agonists but are not primarily responsible for the action; known as guiding muscles, they assist in refined movement and rule out undesired motions.

Neutralizers

Counteract or neutralize the action of another muscle to prevent undesirable movements; referred to as neutralizing, they contract to resist specific actions of other muscles.

When a muscle with multiple agonist actions contracts, it attempts to perform all of its actions. Muscles cannot determine which of their actions are appropriate for the task at hand. The actions actually performed depend on several factors, such as the motor units activated, joint position, muscle length, and relative contraction or relaxation of other muscles acting on the joint. In certain instances, two muscles may work in synergy by counteracting their opposing actions to accomplish a common action.

As discussed, agonist muscles are primarily responsible for a given movement such as that of hip flexion and knee extension during kicking a ball. In this example, the hamstrings are antagonistic and relax to allow the kick to occur. This does not mean all other muscles in the hip area are uninvolved. The preciseness of the kick depends on the involvement of many other muscles. As the lower extremity swings forward, its route and subsequent angle at the point of contact depend on a certain amount of relative contraction or relaxation in the hip abductors, adductors, internal rotators, and external rotators. These muscles act in a synergistic fashion to guide the lower extremity in a precise manner. That is, they are not primarily responsible for knee extension and hip flexion but contribute to the accuracy of the total movement. These guiding muscles assist in refining the kick and preventing extraneous motions. Additionally, the muscles in the contralateral hip and pelvic area must be under relative tension to help fixate or stabilize the pelvis on that side in order to provide a relatively stable pelvis for the hip flexors on the involved side to contract against. When the ball is kicked, the pectineus and tensor fascia latae are adductors and abductors, respectively, in addition to flexors. The actions of abduction and adduction are neutralized by each other, and the common action of the two muscles results in hip flexion.

From a practical point of view, it is not essential that individuals know the exact force exerted by each of the elbow flexors—biceps, brachialis, and brachioradialis—in chinning. It is important to understand that this muscle group is the agonist or primary mover responsible for elbow joint flexion. Similarly, it is important to understand that these muscles contract concentrically when the chin is pulled up to the bar and that they contract eccentrically when the body is lowered slowly. Antagonistic muscles produce actions opposite those of the agonist. For example, the muscles that produce extension of the elbow joint are antagonistic to the muscles that produce flexion of the elbow joint. It is important to understand that specific exercises need to be given for the development of each antagonistic muscle group. The return movement to the hanging position at the elbow joint after chinning is elbow joint extension, but the triceps and anconeus are not being strengthened. A concentric contraction of the elbow joint flexors occurs, followed by an eccentric contraction of the same muscles.

Fig. 1.14, *A* illustrates how the biceps is an agonist by contracting concentrically to flex the elbow. The triceps is an antagonist to elbow flexion, and the pronator teres is considered to be a synergist to the biceps in this example. If the biceps were to slowly lengthen and control elbow extension, it would still be the agonist, but it would be contracting eccentrically. Fig. 1.14, *B* illustrates how the triceps is an agonist by contracting concentrically to extend the elbow. The biceps is an antagonist to elbow extension in this example. If the triceps were to slowly lengthen and control elbow flexion, it would still be the agonist, but it would be eccentrically contracting. In both of these examples, the deltoid, trapezius, and various other shoulder muscles are serving as stabilizers of the shoulder area.

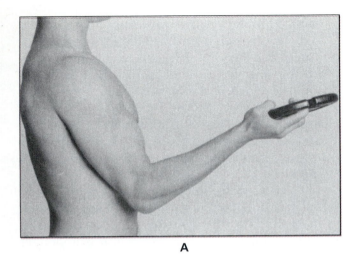

A

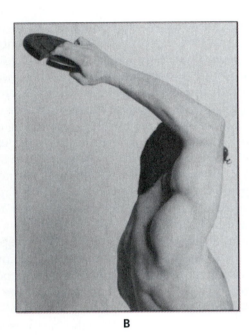

B

FIG. 1.14 ● Agonist-antagonist relationship. **A**, Biceps agonist in elbow flexion; **B**, triceps agonist in elbow extension.

Neuromuscular concepts

All or none principle

When a particular muscle contracts, the contraction actually occurs at the muscle fiber level within a particular motor unit. In a typical muscle contraction, the number of muscle fibers contracting within the muscle may vary significantly from relatively few to virtually all of the muscle fibers, depending on the number of muscle fibers within each activated motor unit and the number of motor units activated. Regardless of the number involved, the individual muscle fibers within a given motor unit will either fire and contract maximally or not at all. This is referred to as the all or none principle. The difference between a particular muscle contracting to lift a minimal versus a maximal resistance is the number of muscle fibers recruited. The number of muscle fibers recruited may be increased by activating those motor units containing a greater number of muscle fibers, by activating more motor units, or by increasing the frequency of motor unit activation.

Muscle length–tension relationship

The maximal ability of a muscle to develop tension and exert force varies, depending on the length of the muscle during contraction. Generally, depending on the particular muscle involved, the greatest amount of tension can be developed when a muscle is stretched between 100% and 130% of its resting length. As a muscle is stretched beyond this point, the amount of force it can exert significantly decreases. Likewise, a proportional decrease in the ability to develop tension occurs as a muscle is shortened. When a muscle is shortened to around 50% to 60% of resting length, its ability to develop contractile tension is essentially reduced to zero. This principle may be seen at work when we squat slightly to stretch the calf, hamstrings, and quadriceps before contracting them concentrically to jump. When attempting to isolate a muscle, we may take advantage of this principle to effectively reduce the contribution of other muscles in the group. For example, in hip extension, we may isolate the work of the gluteus maximus by maximally shortening the hamstrings with flexion of the knee to reduce their ability to act as hip extensors.

Biarticular and multiarticular muscles

Biarticular muscles are those that cross and act on two different joints. Depending on a variety of factors, biarticular muscles may contract and cause motion at either one or both of its joints.

Biarticular muscles have two advantages over uniarticular muscles. They can cause and/or control motion at more than one joint, and they may be able to maintain a relatively constant length due to "shortening" at one joint the "lengthening" at another joint. The muscles does not actually shorten at one joint and lengthen at the other; instead, the concentric shortening of the muscle to move one joint is offset by motion of the other joint which moves its attachment of the muscle farther away. This maintenance of a relatively constant length results in the muscle being able to continue its exertion of force.

The biarticular muscles of the hip and knee provide excellent examples of two different patterns of action. An example of a *concurrent* movement pattern occurs when both the knee and hip extend at the same time. If the knee were to extend only, the rectus femoris would shorten and lose tension, as do the other quadriceps muscles, but its relative length and subsequent tension may be maintained due to its relative lengthening at the hip joint during extension. When a ball is kicked, an example of a *countercurrent* movement pattern may be observed. During the forward movement phase of the lower extremity, the rectus femoris is concentrically contracted to both flex the hip and extend the knee. These two movements when combined increase the tension or stretch on the hamstring muscles both at the knee and hip.

Multiarticular muscles act on three or more joints due to the line of pull between their origin and insertion crossing multiple joints. The principles discussed relative to biarticular muscles apply in a similar fashion to multiarticular muscles.

Reciprocal inhibition or innervation

As stated earlier, antagonist muscles groups must relax and lengthen when the agonist muscle group contracts. This effect, reciprocal innervation, occurs through reciprocal inhibition of the antagonists. Activation of the motor units of the agonists causes a reciprocal neural inhibition of the motor units of the antagonists. This reduction in neural activity of the antagonists allows them to subsequently lengthen under less tension. This may be demonstrated by comparing the ease at which one can stretch the hamstrings when simultaneously contracting the quadriceps versus attempting to stretch the hamstrings without the quadriceps contracted.

Active and passive insufficiency

As a muscle shortens, its ability to exert force diminishes as discussed earlier. When the muscle becomes shortened to the point that it cannot generate or maintain active tension, *active insufficiency* is reached. If the opposing muscle becomes stretched to the point where it can no longer lengthen and allow movement, *passive insufficiency* is reached. These principles are most easily observed in either biarticular or multiarticular muscles when the full range of motion is attempted in all of the joints crossed by the muscle.

An example occurs when the rectus femoris contracts concentrically to both flex the hip and extend the knee. It may completely perform either action one at a time but is actively insufficient to obtain full range at both joints simultaneously. Likewise, the hamstrings will not usually stretch enough to allow both maximal hip flexion and maximal knee extension—hence, they are passively insufficient. As a result of these phenomena, it is virtually impossible to actively extend the knee fully when beginning with the hip fully flexed or vice versa.

Web sites

Anatomy & Physiology Tutorials:

www.gwc.maricopa.edu/class/bio201/index.htm

Neuromusculoskeletal web:

www.sohp.soton.ac.uk/nms/Default.htm

Notes on musculoskeletal anatomy and other links

Kinesiology:

www.kinesiology.org

General information about the field of kinesiology

University of Arkansas Medical School Gross Anatomy for Medical Students:

anatomy.uams.edu/htmlpages/anatomyhtml/gross.html

Dissections, anatomy tables, atlas images, links, etc.

Dynamic Human version 2.0 CD-ROM: The Visual Guide to Anatomy & Physiology:

www.mhhe.com/biosci/ap/dynamichuman2/

Web site that accompanies this CD-ROM

Dynamic Human CD activities

1. Review planes and directional terminology by clicking on **human body**, then **explorations**, and then **anatomical orientation**.
2. Generally review the muscular and skeletal system by clicking on **human body**, then **explorations**, and then **visible human**.
3. Review cartilaginous joints, fibrous joints, long bone cross section, synovial joints, and walking skeleton. Click on **skeletal**, then **explorations**, and then the individual topics listed.
4. Review the skeletal anatomy by clicking on **skeletal** first and then **gross anatomy**.
5. Take the skeletal quiz. Access the quiz by clicking on **skeletal** and then **skeletal quiz**.
6. Explore the muscular anatomy of each region by clicking on **muscular**, then **anatomy**, and then **body regions**.
7. Review the physiology of skeletal muscle by clicking on **muscular**, then **anatomy**, and then **skeletal muscle**.
8. Review each of the following: isometric/isotonic contraction, muscle action around joints, neuromuscular junction, and sliding filament theory by clicking on **muscular**, then **explorations**, and then the individual topic listed.
9. Take the muscular quiz by clicking on **muscular** and then **muscular quiz**.

Worksheet exercises

As an aid to learning, for in-class or out-of-class assignments or for testing, tear-out worksheets are found at the end of the text (see pp. 250 and 251).

Posterior skeletal worksheet (no. 1)

On the posterior skeletal worksheet, list the names of the bones and all of the prominent features of each bone.

Anterior skeletal worksheet (no. 2)

On the anterior skeletal worksheet, list the names of the bones and all of the prominent features of each bone.

Laboratory and review exercises

1. Observe on a fellow student some of the muscles found in Figs. 1.2 and 1.3.
2. Locate the various types of joints on a human skeleton and palpate their movements on a living subject.
3. Individually practice the various joint movements, on yourself or with another subject.
4. The specific body area joint movement terms arise from the basic motions in the three specific planes—flexion/extension in the sagittal plane, abduction/adduction in the frontal plane, and rotation in the transverse plane. With this in mind, complete the joint movement terminology chart on page 23.
5. Determine which joints have movements possible in each of the following planes:
   a. Sagittal
   b. Frontal
   c. Transverse
6. Determine the planes in which the following activities occur. Also, use a pencil to visualize the axis for each of the following activities:
   a. Walking up stairs
   b. Turning a knob to open a door
   c. Nodding the head to agree
   d. Shaking the head to disagree
   e. Shuffling the body from side to side
   f. Looking over your shoulder to see behind
7. Choose several different locations at random on your body and specifically describe the locations, using the correct anatomical directional terminology.
8. Stand facing a closed door. Reach out and grasp the knob. Turn it and open the door widely toward you. Determine all of the joints involved in this activity and list the movements for each joint.
9. With a partner, choose a diarthrodial joint on the body and carry out each of the following exercises:
   a. Familiarize yourself with all of the joint's various movements and list them
   b. Determine which muscle or muscle groups are responsible for each of the movements you listed in 9a
   c. For the muscles or muscle groups you listed for each movement in 9b, determine the type of contraction occurring

d. Determine how to change the parameters of gravity and/or resistance so that the opposite muscles contract to control the same movements in 9c. Name the type of contraction occurring

e. Determine how to change the parameters of movement, gravity, and/or resistance so that the same muscles listed in 9c contract differently to control the opposite movement

10. Complete the joint type, movement, and plane with axes of motion chart by
    a. Filling in the type of diarthrodial joint
    b. Listing the movements of the joint under the plane of motion in which they occur
    c. List the axis of each movement in parentheses immediately following the movement

Joint movement terminology chart

For each specific motion in the left column, provide the basic motion that it represents in the right column by using *flexion, extension, abduction, adduction,* or *rotation* (*external* or *internal*).

Specific motion	Basic motion
Eversion	
Inversion	
Dorsal flexion	
Plantar flexion	
Pronation	
Supination	
Lateral flexion	
Radial flexion	
Ulnar flexion	

Joint type, movement, and plane with axes of motion chart

Joint	Type	Planes of motion		
		Sagittal	Frontal	Transverse
Scapulothoracic joint				
Glenohumeral joint				
Elbow				
Radioulnar joint				
Wrist				
Metacarpophalangeal and metatarsophalangeal joints				
Proximal interphalangeal joints				
Distal interphalangeal joints				
Cervical spine				
Lumbar spine				
Hip				
Knee				
Ankle				

References

Anthony C, Thibodeau G: *Textbook of anatomy and physiology,* ed 10, St. Louis, 1979, Mosby.

Booher JM, Thibodeau GA: *Athletic injury assessment,* ed 4, Dubuque, IA, 2000, McGraw-Hill.

Goss CM: *Gray's anatomy of the human body,* ed 29, Philadelphia, 1973, Lea & Febiger.

Kreighbaum E, Barthels, KM: *Biomechanics: a qualitative approach for studying human movement,* ed 4, Boston, 1996, Allyn & Bacon.

Lindsay DT: *Functional human anatomy,* St. Louis, 1996, Mosby.

Luttgens K, Hamilton N: *Kinesiology: scientific basis of human motion,* ed 9, Madison, WI, 1997, Brown & Benchmark.

Norkin CC, Levangie PK: *Joint structure and function—a comprehensive analysis,* Philadelphia, 1983, Davis.

Northrip JW, Logan GA, McKinney WC: *Analysis of sport motion: anatomic and biomechanic perspectives,* ed 3, 1983, McGraw-Hill Companies, Inc., New York.

Prentice WE: *Rehabilitation techniques in sports medicine,* ed 3, 1999, McGraw-Hill Companies, Inc., New York.

Rasch PJ: *Kinesiology and applied anatomy,* ed 7, Philadelphia, 1989, Lea & Febiger.

Smith LK, Weiss EL, Lehmkuhl LD: *Brunnstrom's clinical kinesiology,* ed 5, Philadelphia, 1996, Davis.

Stedman TL: *Stedman's medical dictionary,* ed 23, Baltimore, 1976, Williams & Wilkins.

Steindler A: *Kinesiology of the human body,* Springfield, IL, 1970, Charles C Thomas.

The shoulder girdle

2

Objectives

- To identify on the skeleton important bone features of the shoulder girdle

- To label on a skeletal chart the important bone features of the shoulder girdle

- To draw on a skeletal chart the muscles of the shoulder girdle and indicate, using arrows, shoulder girdle movements

- To demonstrate, using a human subject, all of the movements of the shoulder girdle and list their respective planes of movement and axes of rotation

- To palpate the muscles of the shoulder girdle on a human subject and list their antagonists

- To palpate the joints of the shoulder girdle on a human subject during each movement through the full range of motion

Brief descriptions of the most important bones in the shoulder region will help you understand the skeletal structure and its relationship to the muscular system.

Bones

Two bones are primarily involved in movements of the shoulder girdle (Figs. 2.1 and 2.2). They are the scapula and clavicle, which generally move as a unit. Their only bony link to the axial skeleton is provided by the clavicle's articulation with the sternum.

Joints

When analyzing shoulder girdle (scapulothoracic) movements, it is important to realize that the scapula moves on the rib cage because the joint motion actually occurs at the sternoclavicular joint and to a lesser amount at the acromioclavicular joint (see Figs. 2.1 and 2.2).

Sternoclavicular (SC): Classified as a (multiaxial) arthrodial joint. It moves anteriorly 15 degrees with protraction and posteriorly 15 degrees with retraction. It moves superiorly 45 degrees with elevation and inferiorly 5 degrees with depression. It is supported anteriorly by the anterior sternoclavicular ligament and posteriorly by the posterior ligament. Additionally, the costoclavicular and interclavicular ligaments provide stability against superior displacement.

Acromioclavicular (AC): Classified as an arthrodial joint. It has a 20- to 30-degree total gliding and rotational motion accompanying other shoulder girdle and shoulder joint motions. In addition to the strong support provided by the coracoclavicular ligaments (trapezoid and conoid), the superior and inferior acromioclavicular ligaments provide stability to this often injured joint. The coracoclavicular joint, classified as a syndesmotic type joint, functions through its ligaments to greatly increase the stability of the acromioclavicular joint.

Scapulothoracic: Not a true synovial joint, due to its not having regular synovial features and due fact that its movement is totally dependent on the sternoclavicular and acromioclavicular joints. Even though scapula movement occurs as a result of motion at the SC and AC joints, the scapula can be described as having a total range of 25-degree abduction-adduction movement, 60-degree upward-downward rotation, and 55-degree elevation-depression. The scapulothoracic joint is supported dynamically by its muscles and lacks ligamentous support, since it has no synovial features.

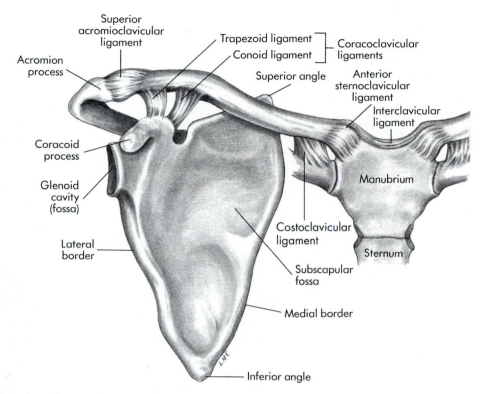

FIG. 2.1 ● Right shoulder girdle, anterior view.

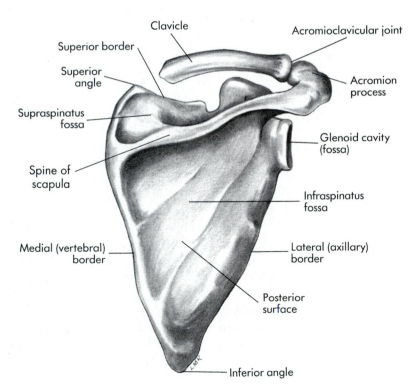

FIG. 2.2 ● Right scapula, posterior view.

Movements FIG. 2.3 & FIG. 2.4

In analyzing shoulder girdle movements, it is often helpful to focus on a specific scapular bony landmark, such as the inferior angle (posteriorly), the glenoid fossa (laterally), and the acromion process (anteriorly). All of these movements have their pivotal point where the clavicle joins the sternum at the sternoclavicular joint.

Movements of the shoulder girdle can be described as movements of the scapula.

Abduction (protraction): movement of the scapula laterally away from the spinal column

Adduction (retraction): movement of the scapula medially toward the spinal column

Upward rotation: turning the glenoid fossa upward and moving the inferior angle superiorly and laterally away from the spinal column

Downward rotation: returning the inferior angle medially and inferiorly toward the spinal column and the glenoid fossa to its normal position.

Elevation: upward or superior movement, as in shrugging the shoulders

Depression: downward or inferior movement, as in returning to normal position

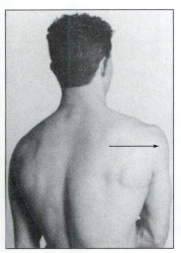

Abduction
(protraction)

A

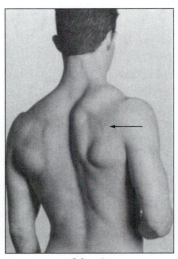

Adduction
(retraction)

B

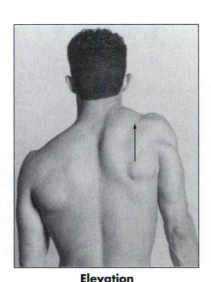

Elevation

C

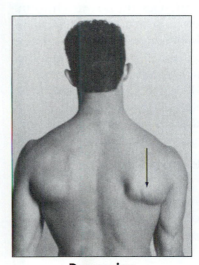

Depression

D

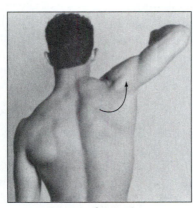

Upward rotation

E

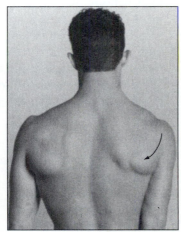

Downward rotation

F

FIG. 2.3 ● Movements of the shoulder girdle.

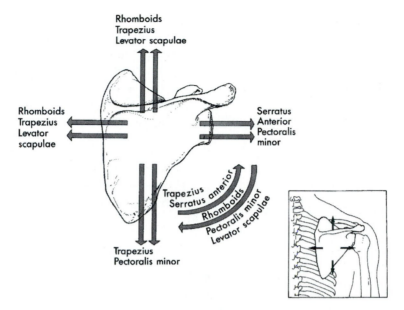

FIG. 2.4 ● Actions of the scapular muscles.

The shoulder joint and shoulder girdle work together in carrying out upper extremity activities. It is critical to understand that movement of the shoulder girdle is not dependent on the shoulder joint and its muscles. However, the muscles of the shoulder girdle are essential in providing a scapula stabilizing effect, so that the muscles of the shoulder joint will have a stable base from which to exert force for powerful movement involving the humerus. Consequently, the shoulder girdle muscles contract to maintain the scapula in a relatively static position during many shoulder joint actions. As the shoulder joint goes through more extreme ranges of motion, the scapular muscles contract to move the shoulder girdle as well to enhance movement of the entire upper extremity. The shoulder girdle movements that usually accompany shoulder joint movements are addressed in Table 3.1 in Chapter 3.

Muscles

There are five muscles primarily involved in shoulder girdle movements. To avoid confusion, it is helpful to group muscles of the shoulder girdle separately from the shoulder joint. The subclavius muscle is also included in this group, but it is not regarded as a primary mover in any actions of the shoulder girdle. All five shoulder girdle muscles have their origin on the axial skeleton, with their insertion located on the scapula and/or the clavicle. Shoulder girdle muscles do not attach to the humerus, nor do they cause actions of the shoulder joint. The pectoralis minor, serratus anterior, and subclavius are located anteriorly, and the trapezius, rhomboid, and levator scapula are posterior.

The shoulder girdle muscles are essential in providing dynamic stability of the scapula, so that it can serve as a relative base of support for shoulder joint activities such as throwing, batting, and blocking.

Shoulder girdle muscle—location

Anterior
    Primarily abduction and depression
        Pectoralis minor
        Serratus anterior
        Subclavius
Posterior
    Primarily adduction and elevation
        Trapezius
        Rhomboid
        Levator scapula

Trapezius muscle   FIG. 2.5

(tra-pe´zi-us)

Origin

Upper fibers: base of skull, occipital protuberance, and posterior ligaments of neck

Middle fibers: spinous processes of seventh cervical and upper three thoracic vertebrae

Lower fibers: spinous processes of fourth through twelfth thoracic vertebrae

Insertion

Upper fibers: posterior aspect of the lateral third of the clavicle

Middle fibers: medial border of the acromion process and upper border of the scapular spine

Lower fibers: triangular space at the base of the scapular spine

Action

Upper fibers: elevation of the scapula; extension of the head at the neck

Middle fibers: elevation, upward rotation, and adduction of the scapula

Lower fibers: depression, adduction, and upward rotation of the scapula

Palpation

Large area up and down from the neck region to the twelfth thoracic spine and laterally from the vertebral column to the scapula

Innervation

Accessory nerve (cranial nerve XI) and branches of C3, 4

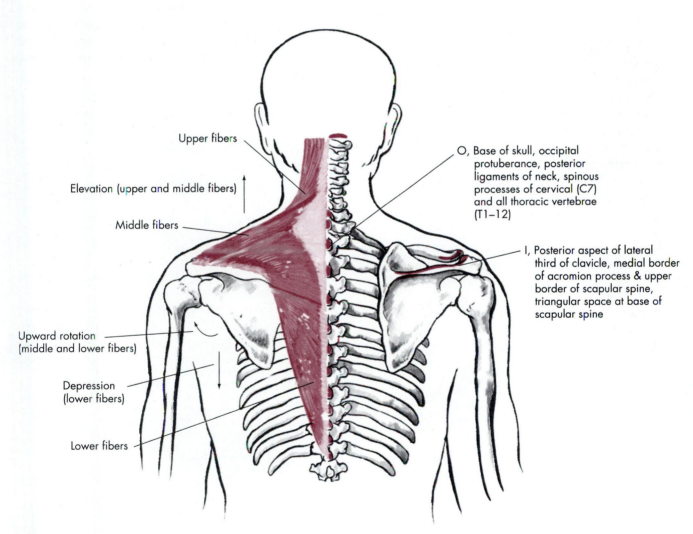

Upper fibers

Elevation (upper and middle fibers)

Middle fibers

Upward rotation (middle and lower fibers)

Depression (lower fibers)

Lower fibers

O, Base of skull, occipital protuberance, posterior ligaments of neck, spinous processes of cervical (C7) and all thoracic vertebrae (T1–12)

I, Posterior aspect of lateral third of clavicle, medial border of acromion process & upper border of scapular spine, triangular space at base of scapular spine

FIG. 2.5 ● Trapezius muscle. *O,* Origin; *I,* insertion; posterior view.

Application, strengthening, and flexibility

The upper fibers are a thin and relatively weak part of the muscle. They provide some elevation of the clavicle. As a mover of the head, they are of minor importance.

The middle fibers are stronger and thicker and provide strong elevation, upward rotation, and adduction (retraction) of the scapula.

The lower fibers assist in adduction (retraction) and rotate the scapula upward.

When all the parts of the trapezius are working together, they tend to pull upward and adduct at the same time. Typical action of the trapezius muscle is fixation of the scapula for deltoid action. Continuous action in upward rotation of the scapula permits the arms to be raised over the head. The muscle is always used in preventing the glenoid fossae from being pulled down during the lifting of objects with the arms. It is also typically seen in action during the holding of an object overhead. Holding the arm at the side horizontally shows typical fixation of the scapula by the trapezius muscle, while the deltoid muscle holds the arm in that position. The muscle is used strenuously when lifting with the hands, as in picking up a heavy wheelbarrow. The trapezius must prevent the scapula from being pulled downward. Carrying objects on the tip of the shoulder also calls this muscle into play. Strengthening of the upper and middle fibers can be accomplished through shoulder-shrugging exercises. The middle and lower fibers can be strengthened through bent rowing from a prone position and side arm shoulder joint abduction exercises.

To stretch the trapezius, each portion needs to be specifically addressed. The upper fibers may be stretched by using one hand to pull the head and neck forward into flexion or slight lateral flexion to the opposite side while the ipsilateral hand is hooked under a table edge to maintain the scapula in depression. The middle fibers are stretched to some extent with the procedure used for the upper fibers, but they may be stretched further by using a partner to passively pull the scapula into full protraction. The lower fibers are perhaps best stretched with the subject in a side-lying position while a partner grasps the lateral border and inferior angle of the scapula and moves it passively into maximal elevation and protraction.

Levator scapulae muscle   FIG. 2.6

(le-va´tor scap´u-lae)

Origin

Transverse processes of the upper four cervical vertebrae

Insertion

Medial border of the scapula above the base of the scapular spine

Action

Elevates the medial margin of the scapula

Palpation

Cannot be palpated; deep to the trapezius muscle

Innervation

Dorsal scapula nerve C5 and branches of C3 and C4

Application, strengthening, and flexibility

Shrugging the shoulders calls the levator scapulae muscle into play, along with the upper trapezius muscle. Fixation of the scapula by the pectoralis minor muscle allows the levator scapulae muscles on both sides to extend the neck or to flex laterally if used on one side only.

The levator scapula is perhaps best stretched by rotating the head approximately 45 degrees to the opposite side and flexing the cervical spine actively while maintaining the scapula in a relaxed, depressed position.

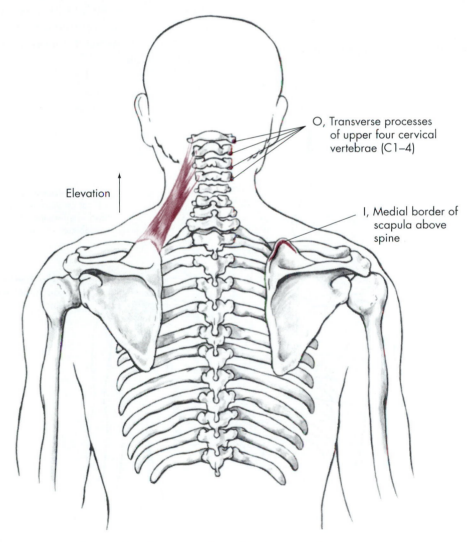

Elevation

O, Transverse processes of upper four cervical vertebrae (C1–4)

I, Medial border of scapula above spine

FIG. 2.6 ● Levator scapulae muscle, *O*, Origin; *I*, insertion; posterior view.

Rhomboid muscles—major and minor FIG. 2.7

(rom´boyd)

Origin

Spinous processes of the last cervical and the first five thoracic vertebrae

Insertion

Medial border of the scapula, below the spine

Action

The rhomboid major and minor muscles work together

Adduction (retraction): draw the scapula toward the spinal column

Rotation downward: from the upward rotated position; they draw the scapula into downward rotation

Elevation: slight upward movement accompanying adduction

Palpation

Cannot be palpated; deep to the trapezius muscle

Innervation

Dorsal scapula nerve (C5)

Application, strengthening, and flexibility

The rhomboid muscles fix the scapula in adduction (retraction) when the muscles of the shoulder joint adduct or extend the arm. These muscles are used powerfully in chinning. As one hangs from the horizontal bar, suspended by the hands, the scapula tends to be pulled away from the top of the chest. When the chinning movement begins, it is the rhomboid muscles that rotate the medial border of the scapula down and back toward the spinal column. Note their favorable position to do this.

The trapezius and rhomboid muscles working together produce adduction with slight elevation of the scapula. To prevent this elevation, the latissimus dorsi muscle is called into play.

Chin-ups and dips are excellent exercises for developing strength in this muscle. The rhomboids may be stretched by passively moving the scapula into full protraction while maintaining depression. Upward rotation may assist in this stretch as well.

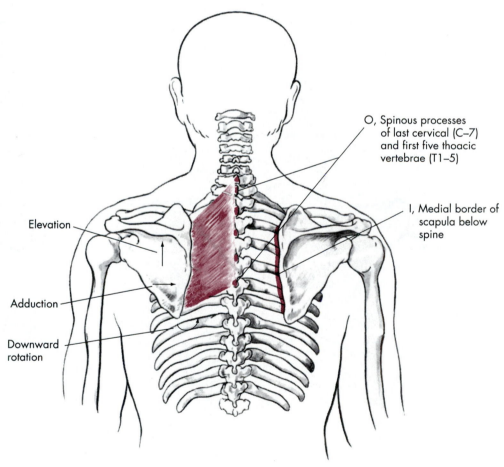

Elevation

Adduction

Downward rotation

O, Spinous processes of last cervical (C–7) and first five thoacic vertebrae (T1–5)

I, Medial border of scapula below spine

FIG. 2.7 ● Rhomboid muscles (major and minor). *O,* Origin; *I,* insertion; posterior view.

Serratus anterior muscle   FIG. 2.8

(ser-a´tus an-tir´e-or)

Origin

Surface of the upper nine ribs at the side of the chest

Insertion

Anterior aspect of the whole length of the medial
  border of the scapula

Action

Abduction (protraction): draws the medial border of
  the scapula away from the vertebrae

Rotation upward: longer, lower fibers tend to draw the
  inferior angle of the scapula farther away from the
  vertebrae, thus rotating the scapula upward slightly

Palpation

Front and lateral side of the chest below the fifth
  and sixth ribs

Innervation

Long thoracic nerve (C5–7)

Application, strengthening, and flexibility

The serratus anterior muscle is used commonly in
movements drawing the scapula forward with
slight upward rotation, such as throwing a baseball, shooting and guarding in basketball, and
tackling in football. It works along with the pectoralis major muscle in typical action, such as
throwing a baseball.

The serratus anterior muscle is used strongly in
doing push-ups, especially in the last 5 to 10 degrees of motion. The bench press and overhead
press are good exercises for this muscle. A
winged scapula condition indicates a definite
weakness of the serratus anterior, which may result from an injury to the long thoracic nerve.

The serratus anterior can be stretched by standing, facing a corner and placing each hand at
shoulder level on the two walls. As you lean in
and attempt to place your nose in the corner, both
scapula are pushed into an adducted position,
which stretches the serratus anterior.

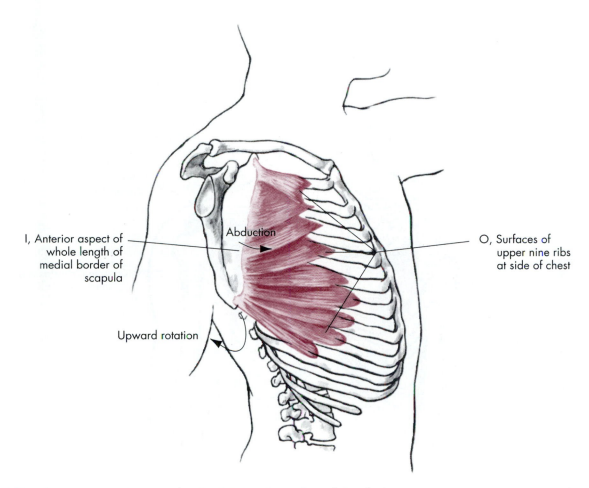

I, Anterior aspect of whole length of medial border of scapula

Abduction

O, Surfaces of upper nine ribs at side of chest

Upward rotation

FIG. 2.8 ● Serratus anterior muscle. *O,* Origin; *I,* insertion; lateral view.

Pectoralis minor muscle FIG. 2.9

(pek-to-ra´lis mi´nor)

Origin

Anterior surfaces of the third to fifth ribs

Insertion

Coracoid process of the scapula

Action

Abduction (protraction): draws the scapula
  forward and tends to tilt the lower border away
  from the ribs

Downward rotation: as it abducts, it draws the
  scapula downward

Depression: when the scapula is rotated upward, it
  assists in depression

Palpation

Difficult to palpate, but can be palpated under the
  pectoralis major muscle in the pit of the shoulder
  during powerful downward movement

Innervation

Medial pectoral nerve (C8–T1)

Application, strengthening, and flexibility

The pectoralis minor muscle is used, along with
the serratus anterior muscle, in true abduction
(protraction) without rotation. This is seen particularly in movements such as push-ups. True
abduction of the scapula is necessary. Therefore,
the serratus anterior draws the scapula forward
with a tendency toward upward rotation, the
pectoralis minor pulls forward with a tendency
toward downward rotation, and the two pulling
together give true abduction, which is necessary
in push-ups. These muscles will be seen working
together in most movements of pushing with the
hands.

The pectoralis minor is most used in depressing and rotating the scapula downward from an
upwardly rotation position, such as in pushing the
body upward on dip bars.

The wall push-up in the corner, as used for
the serratus anterior, is helpful for improving
flexibility in the pectoralis minor. Additionally,
lying supine with a rolled towel directly under the
thoracic spine while a partner pushes each
scapula into retraction places this muscle on
stretch.

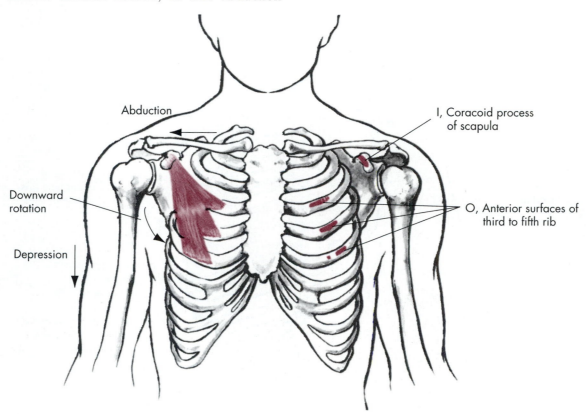

FIG. 2.9 ● Pectoralis minor muscle. *O,* Origin; *I,* insertion; anterior view.

Subclavius muscle   FIG. 2.10

(sub-klá-ve-us)

Origin

Superior aspect of first rib at its junction with its
costal cartilage

Insertion

Inferior groove in the midportion of the clavicle

Action

Stabilization and protection of the sternoclavicular
joint
Depression

Palpation

Cannot be palpated

Innervation

Nerve fibers from C5 and C6

Application, strengthening, and flexibility

The subclavius pulls the clavicle inferiorly toward
the sternum. In addition to assisting in depressing
the clavicle and the shoulder girdle, it has a significant role in protecting and stabilizing the sternoclavicular joint during upper extremity movements. It may be strengthened during activities in
which there is active depression, such as dips. Extreme elevation of the shoulder girdle provides a
stretch to the subclavius.

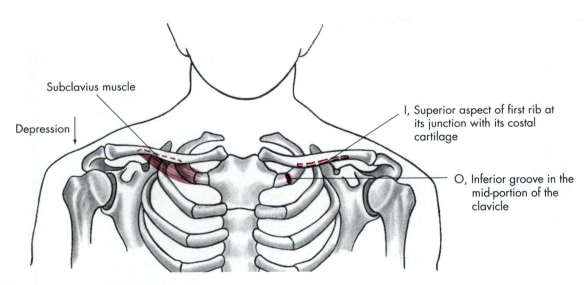

FIG. 2.10 ● Subclavius muscle. *O*, Origin; *I*, insertion; anterior view.

Web sites

Anatomy & Physiology Tutorials:

> **www.gwc.maricopa.edu/class/bio201/index.htm**

Electronic Textbook of Hand Surgery:

> **http://www.e-hand.com/default.htm**
>
> Slides and illustrations of upper extremity musculoskeletal anatomy

Radiologic Anatomy Browser:

> **radlinux1.usuf1.usuhs.mil/rad/iong/index.html**
>
> This site has numerous radiological views of the musculoskeletal system.

University of Arkansas Medical School Gross Anatomy for Medical Students:

> **http://anatomy.uams.edu/htmlpages/anatomyhtml/gross.html**
>
> Dissections, anatomy tables, atlas images, links, etc.

Loyola University Medical Center: Structure of the Human Body:

> **www.meddean.luc.edu/lumen/MedEd/GrossAnatomy/GA.html**
>
> An excellent site with many slides, dissections, tutorials, etc., for the study of human anatomy

Wheeless' Textbook of Orthopaedics:

> **www.medmedia.com/**
>
> This site has an extensive index of links to the fractures, joints, muscles, nerves, trauma, medications, medical topics, lab tests, and links to orthopedic journals and other orthopedic and medical news.

Anatomy of the Shoulder Tutorial:

> **www.ncl.ac.uk/~nccc/tutorials/shoulder/**

Premiere Medical Search Engine:

> **www.medsite.com**
>
> This site allows the reader to enter any medical condition and it will search the net to find relevant articles.

Virtual Hospital:

> **www.vh.org**
>
> Numerous slides, patient information, etc.

Dynamic Human version 2.0 CD-ROM: The Visual Guide to Anatomy & Physiology:

> **www.mhhe.com/biosci/ap/dynamichuman2/**
>
> Web site that accompanies the CD-ROM

Dynamic Human CD activities

1. Review anatomical landmarks as well as origins/insertions for the clavicle and scapula by clicking on **skeletal**; then **anatomy**; then **gross anatomy, pectoral girdle,** and **clavicle**; and then **scapula**.
2. Review each of the muscles from this chapter by clicking on **muscular**; then **anatomy**; then **body regions, pectoral girdle,** and **upper arm**; and then **head and neck**.

Worksheet exercises

As an aid to learning, for in-class or out-of-class assignments, or for testing, tear-out worksheets are found at the end of the text (pp. 252 and 253).

Skeletal worksheet (no. 1)

Draw and label on the worksheet the following muscles:

a. Trapezius
b. Rhomboid major and minor
c. Serratus anterior
d. Levator scapulae
e. Pectoralis minor
f. Subclavius

Human figure worksheet (no. 2)

Label the circles next to the arrows with the appropriate letter that corresponds to the movements of the shoulder girdle indicated by the arrow:

a. Adduction (retraction)
b. Abduction (protraction)
c. Rotation upward
d. Rotation downward
e. Elevation
f. Depression

Laboratory and review exercises

1. Locate the following prominent skeletal features on a human skeleton and on a subject:
   a. Scapula
      1. Medial border
      2. Inferior angle
      3. Superior angle
      4. Coracoid process
      5. Spine of scapula
      6. Glenoid cavity
      7. Acromion process
      8. Supraspinatus fossa
      9. Infraspinatus fossa

b. Clavicle
1. Sternal end
2. Acromial end
c. Joints
1. Sternoclavicular joint
2. Acromioclavicular joint
2. How and where do you palpate the following muscles on a human subject?
a. Serratus anterior
b. Trapezius
c. Rhomboid major and minor
d. Levator scapulae
e. Pectoralis minor
NOTE: "How" means resisting a primary movement of the muscle. Some muscles have several primary movements, such as the trapezius with rotation upward and adduction. "Where" refers to the location on the body where the muscle can be felt.
3. Palpate the sternoclavicular and acromioclavicular joint movements and the muscles primarily involved while demonstrating the following shoulder girdle movements.
a. Adduction
b. Abduction

c. Rotation upward
d. Rotation downward
e. Elevation
f. Depression
4. List the planes in which each of the following shoulder girdle movements occur. List the respective axis of rotation for each movement in each plane.
a. Adduction
b. Abduction
c. Rotation upward
d. Rotation downward
e. Elevation
f. Depression
5. Fill in the muscle analysis chart by listing the muscles primarily involved in each movement.
6. Fill in the antagonistic muscle action chart by listing the muscle(s) or parts of muscles that are antagonist in their actions to the muscles in the left column.

Muscle analysis chart • Shoulder girdle

Shoulder girdle	
Abduction	Adduction
Elevation	Depression
Upward rotation	Downward

Antagonistic muscle action chart • Shoulder girdle

Agonist	Antagonist
Serratus anterior	
Trapezius (upper fibers)	
Trapezius (middle fibers)	
Trapezius (lower fibers)	
Rhomboid	
Levator scapulae	
Pectoralis minor	

References

Andrews JR, Wilk KE: *The athlete's shoulder,* New York, 1994, Churchill Livingstone.

Andrews JR, Zarins B, Wilk KE: *Injuries in baseball,* Philadelphia, 1998, Lippincott-Raven.

Hislop HJ, Montgomery J: *Daniels and Worthingham's muscle testing: techniques of manual examination,* ed 6, Philadelphia, 1995, Saunders.

McMurtrie H, Rikel JK: *The coloring review guide to human anatomy,* 1991, McGraw-Hill Companies, Inc., New York.

Norkin CC, Levangie PK: *Joint structure and functional comprehensive analysis,* Philadelphia, 1983, Davis.

Rasch PJ: *Kinesiology and applied anatomy,* ed 7, Philadelphia, 1989, Lea & Febiger.

Smith LK, Weiss EL, Lehmkuhl LD: *Brunnstrom's clinical kinesiology,* ed 5, Philadelphia, 1996, Davis.

Sobush DC, et al: The Lennie test for measuring scapula position in healthy young adult females: a reliability and validity study, *Journal of Orthopedic and Sports Physical Therapy* 23:39, January 1996.

Soderburg GL: *Kinesiology—application to pathological motion,* Baltimore, 1986, Williams & Wilkins.

The shoulder joint 3

glenohumeral joint is classified as Enarthrodial

Objectives

- To identify on a human skeleton or human subject selected bony structures of the shoulder joint

- To label on a skeletal chart selected bony structures of the shoulder joint

- To draw on a skeletal chart the muscles of the shoulder joint and indicate, using arrows, shoulder joint movements

- To demonstrate with a fellow student all of the movements of the shoulder joints and list their respective planes and axes of rotation

- To learn and understand how movements of the scapula accompany movements of the humerus in achieving movement of the entire shoulder complex

- To determine and list the muscles of the shoulder joint and their antagonists

- To organize and list the muscles that produce the movements of the shoulder girdle and the shoulder joint

The only attachment of the shoulder joint to the axial skeleton is with the clavicle at the sternoclavicular joint. Movements of the shoulder joint are many and varied. It is unusual to have movement of the humerus without scapula movement. When the humerus is flexed above shoulder level, the scapula is elevated, rotated upward, and abducted. With glenohumeral abduction above shoulder level, the scapula is rotated upward and elevated. Adduction of the humerus results in rotation downward and depression, whereas extension of the humerus results in depression, rotation downward, and adduction of the scapula.

The scapula abducts with humeral internal rotation and horizontal adduction. Scapula adduction accompanies external rotation and horizontal abduction of the humerus. For a summary of these movements, refer to Table 3.1.

Because the shoulder joint has such a wide range of motion in so many different planes, it also has a significant amount of laxity, which often results in instability problems such as rotator cuff impingement, subluxations, and dislocations. The price of mobility is reduced stability. The concept that the more mobile a joint is, the less stable it is and that the more stable it is, the less mobile it is applies generally throughout the body but particularly in the shoulder joint.

Bones

The scapula, clavicle, and humerus serve as attachments for most of the muscles of the shoulder joint. Learning the specific location and importance of certain bony landmarks is critical to understanding the functions of the shoulder complex. Some of these scapular landmarks are the supraspinatus fossa, infraspinatus fossa, subscapular fossa, spine of the scapula, glenoid cavity, coracoid process, acromion process, and inferior angle. Humeral landmarks are the head, greater tubercle, lesser tubercle, intertubercular groove, and deltoid tuberosity (review Fig. 2.1 and see Fig. 3.1).

Joint

The shoulder joint, specifically known as the glenohumeral joint, is a multiaxial ball-and-socket joint classified as enarthrodial (Fig. 3.1). Its stability is enhanced slightly by the glenoid labrum, a cartilaginous ring that surrounds the glenoid fossa just inside its periphery. It is further stabilized by the glenohumeral ligaments, especially anteriorly and inferiorly. The anterior glenohumeral ligaments become taut as external rotation, extension, abduction, and horizontal abduction occur, whereas the very thin posterior capsular ligaments become taut in internal rotation, flexion, and horizontal adduction. In recent years the importance of the inferior glenohumeral ligament in providing both anterior and posterior stability has come to light. It is, however, important to note that, due to the wide range of motion involved in the glenohumeral joint, the ligaments are quite lax until the extreme ranges of motion are reached. Stability is sacrificed to gain mobility.

Movement of the humerus from the side position is common in throwing, tackling, and striking activities. Flexion and extension of the shoulder joint are performed frequently when supporting body weight in a hanging position or in a movement from a prone position on the ground.

Determining the exact range of each movement for the glenohumeral joint is difficult because of the accompanying shoulder girdle movement. However, the glenohumeral joint movements are generally thought to be in the following ranges: 90 to 95 degrees abduction, 0 degrees adduction (prevented by the trunk) or 75 degrees anterior to the trunk, 40 to 60 degrees of extension, 90 to 100 degrees of flexion, 70 to 90 degrees of internal and external rotation, 45 degrees of horizontal abduction, and 135 degrees of horizontal adduction.

The shoulder joint is frequently injured because of its anatomical design. A number of factors contribute to its injury rate, including the shallowness of the glenoid fossa, the laxity of the ligamentous structures necessary to accommodate its wide range of motion, and the lack of strength and endurance in the muscles, which are essential in providing dynamic stability to the joint. As a result, anterior or anteroinferior glenohumeral subluxations and dislocations are quite common with physical activity. Although posterior dislocations are fairly rare in occurrence, problems about the shoulder due to posterior instability are somewhat commonplace.

Another frequent injury is to the rotator cuff. The subscapularis, supraspinatus, infraspinatus, and teres minor muscles make up the rotator cuff. They are small muscles whose tendons cross the front, top, and rear of the head of the humerus to attach on the lesser and greater tuberosities, respectively. Their point of insertion enables them to rotate the humerus, an essential movement in this freely movable joint. Most important, however, is the vital role that the rotator cuff muscles play in maintaining the humeral head in correct approximation within the glenoid fossa while the more powerful muscles of the joint move the humerus through its wide range of motion.

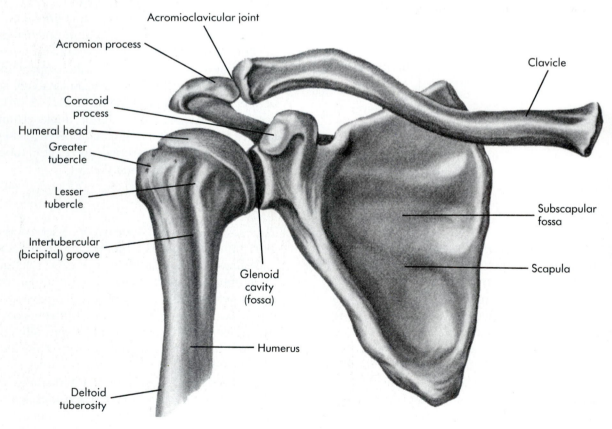

FIG. 3.1 ● Right glenohumeral joint, anterior view.

TABLE 3.1 • Pairing of shoulder girdle and shoulder joint movements

Shoulder joint	Shoulder girdle
Abduction	Upward rotation
Adduction	Downward rotation
Flexion	Elevation/upward rotation
Extension	Depression/downward rotation
Internal rotation	Abduction (protraction)
External rotation	Adduction (retraction)
Horizontal abduction	Adduction (retraction)
Horizontal adduction	Abduction (protraction)
Diagonal abduction (overhand activities)	Adduction (retraction)/upward rotation/elevation
Diagonal abduction (overhand activities)	Abduction (protraction)/depression/downward rotation

Movements FIG. 3.2

Abduction: upward lateral movement of the humerus out to the side, away from the body

Adduction: downward movement of the humerus medially toward the body from abduction

Flexion: movement of the humerus straight anteriorly

Extension: movement of the humerus straight posteriorly, sometimes referred to as hyperextension

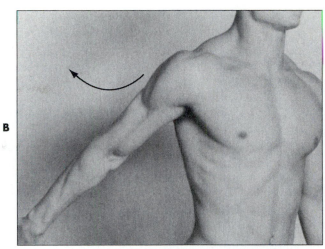

Extension

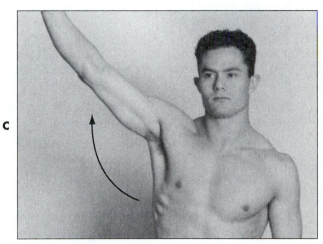

Abduction

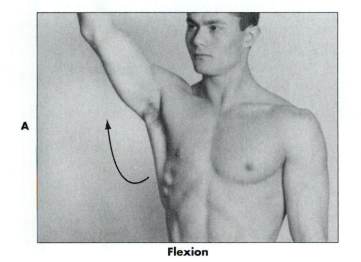

Flexion

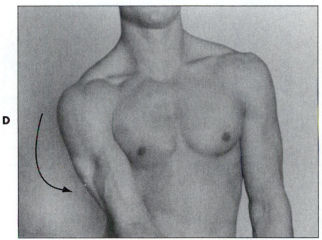

Adduction

FIG. 3.2 • Movements of the shoulder joint.

Horizontal adduction (flexion): movement of the humerus in a horizontal or transverse plane toward and across the chest

Horizontal abduction (extension): movement of the humerus in a horizontal or transverse plane away from the chest

External rotation: movement of the humerus laterally around its long axis away from the midline

Internal rotation: movement of the humerus medially around its long axis toward the midline

Diagonal abduction: movement of the humerus in a diagonal plane away from the midline of the body

Diagonal adduction: movement of the humerus in a diagonal plane toward the midline of the body

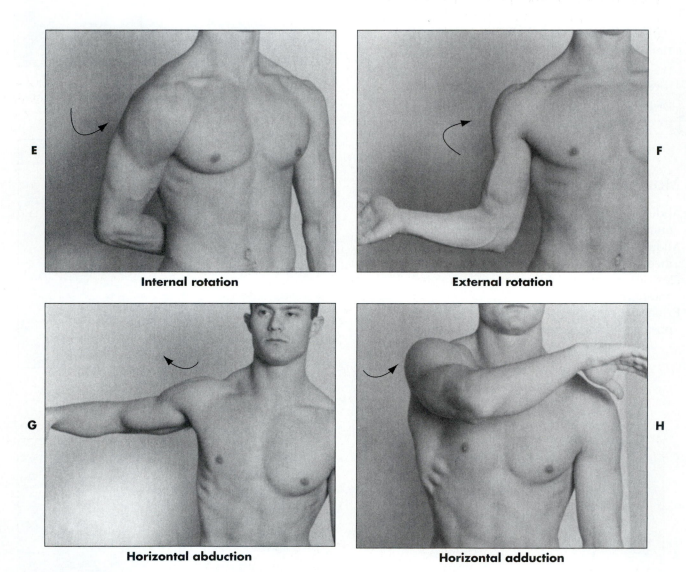

E   **Internal rotation**   F   **External rotation**

G   **Horizontal abduction**   H   **Horizontal adduction**

FIG. 3.2 continued ● Movements of the shoulder joint.

Muscles

When attempting to learn and understand the muscles of the glenohumeral joint, it may be helpful to group them according to their location and function. All of the muscles that originate on the scapula and clavicle may be thought of as intrinsic glenohumeral muscles. The intrinsic muscles include the deltoid, the coracobrachialis, the teres major, and the rotator cuff group, which is composed of the subscapularis, the supraspinatus, the infraspinatus, and the teres minor. Extrinsic glenohumeral muscles are the latissimus dorsi and pectoralis major. It may also be helpful to organize the muscles according to their general location. The pectoralis major, coracobrachialis, and subscapularis are anterior muscles. Located superiorly are the deltoid and supraspinatus. The latissimus dorsi, teres major, infraspinatus, and teres minor are located posteriorly.

The biceps brachii and triceps brachii (long head) are also involved in glenohumeral movements. Primarily, the biceps brachii assists in flexing and horizontally adducting the shoulder, whereas the long head of the triceps brachii assists in extension and horizontal abduction. Further discussion of these muscles appears in Chapter 4.

Shoulder joint muscles—location

Anterior
    Pectoralis major
    Coracobrachialis
    Subscapularis
Superior
    Deltoid
    Supraspinatus
Posterior
    Latissimus dorsi
    Teres major
    Infraspinatus
    Teres minor

Muscle identification

In Figs. 3.3 and 3.5, identify the anterior and posterior muscles of the shoulder joint and shoulder girdle. Compare Fig. 3.3 with Fig. 3.4 and Fig. 3.5 with Fig. 3.6.

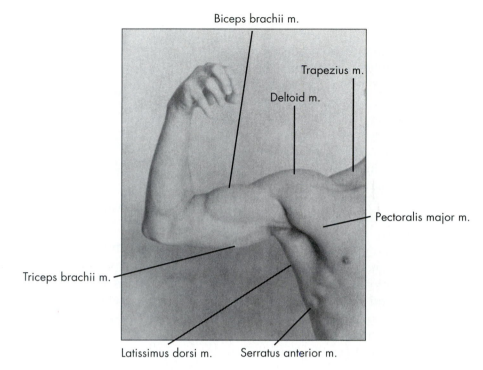

FIG. 3.3 ● Anterior shoulder joint and shoulder girdle muscles.

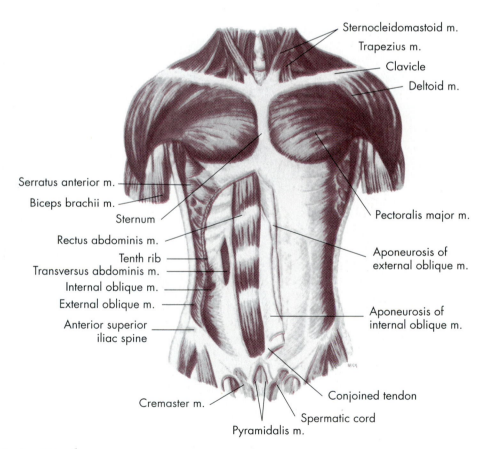

FIG. 3.4 ● Anterior muscles.

From Booher JA, Thibodeau GA: *Athletic injury assessment*, ed 2, St. Louis, 1989, Mosby.

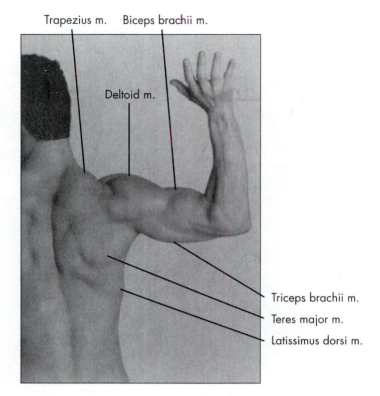

Trapezius m.   Biceps brachii m.

Deltoid m.

Triceps brachii m.

Teres major m.

Latissimus dorsi m.

FIG. 3.5 ● Posterior shoulder joint and shoulder girdle muscles.

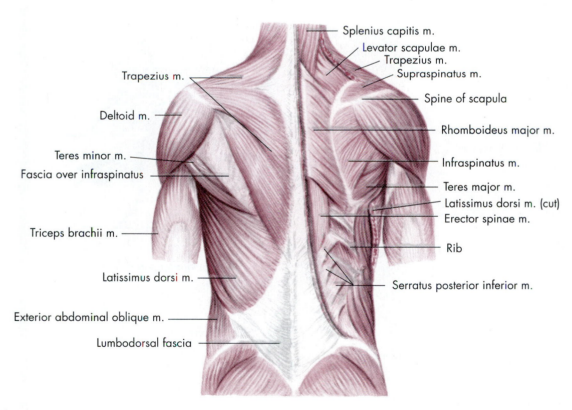

Splenius capitis m.

Levator scapulae m.

Trapezius m.

Supraspinatus m.

Trapezius m.

Spine of scapula

Deltoid m.

Rhomboideus major m.

Teres minor m.

Infraspinatus m.

Fascia over infraspinatus

Teres major m.

Latissimus dorsi m. (cut)

Erector spinae m.

Triceps brachii m.

Rib

Latissimus dorsi m.

Serratus posterior inferior m.

Exterior abdominal oblique m.

Lumbodorsal fascia

FIG. 3.6 ● Posterior muscles.

Rotator cuff muscles

Fig. 3.7 illustrates the rotator cuff muscle group, which, as previously mentioned, is most important in maintaining the humeral head in its proper location within the glenoid cavity. The acronym **SITS** may be used in learning the names of the supraspinatus, infraspinatus, teres minor, and subscapularis. These muscles, which are not very large in comparison with the deltoid and pectoralis major, must possess not only adequate strength but also a significant amount of muscular endurance to ensure their proper functioning, particularly in repetitious overhead activities such as throwing, swimming, and pitching. Quite often when these types of activities are conducted with poor technique, muscle fatigue, or inadequate warm-up and conditioning, the rotator cuff muscle group, particularly the supraspinatus, fails to dynamically stabilize the humeral head in the glenoid cavity, leading to further rotator cuff problems such as tendinitis and rotator cuff impingement within the subacromial space.

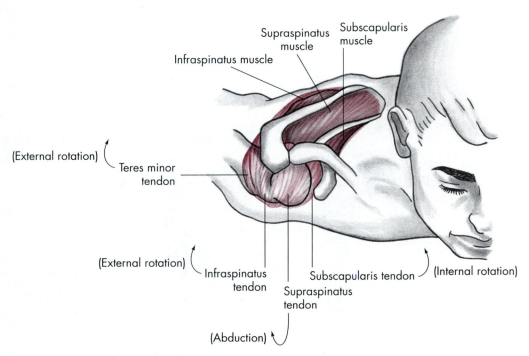

FIG. 3.7 ● Rotator cuff muscles, superior view.

Deltoid muscle   FIG. 3.8

(del-toyd´)

Origin

Anterior fibers: anterior lateral third of the clavicle
Middle fibers: lateral aspect of the acromion
Posterior fibers: inferior edge of the spine of the
  scapula

Insertion

Deltoid tuberosity on the lateral humerus

Action

Anterior fibers: abduction, flexion, horizontal
  adduction, and internal rotation of the
  glenohumeral joint
Middle fibers: abduction of the glenohumeral joint
Posterior fibers: abduction, extension, horizontal
  abduction, and external rotation of the
  glenohumeral joint

Palpation

Over the head of the humerus from the anterior to
  the posterior side

Innervation

Axillary nerve (C5, 6)

Application, strengthening, and flexibility

The deltoid muscle is used commonly in any lifting movement. The trapezius muscle stabilizes the
scapula as the deltoid pulls on the humerus. The
anterior fibers of the deltoid muscle flex and internally rotate the humerus. The posterior fibers
extend and externally rotate the humerus. The anterior fibers also horizontally adduct the humerus
while the posterior fibers horizontally abduct it.

This muscle is used in all lifting movements if
the arms are at the sides in lifting.

Any movement of the humerus on the scapula
will involve part or all of the deltoid muscle.

Lifting the humerus from the side to the position of abduction is a typical action of the deltoid.
Side-arm dumbbell raises are excellent for
strengthening the deltoid, especially the middle
fibers. By abducting the arm in a slightly horizontally adducted (30 degrees) position, the anterior
deltoid fibers can be emphasized. The posterior
fibers can be strengthened better by abducting the
arm in a slightly horizontally abducted (30 degrees) position.

Stretching the deltoid requires varying positions, depending on the fibers to be stretched.
The anterior deltoid is stretched by taking the
humerus into extreme horizontal abduction or by
extreme extension and adduction. The middle
deltoid is stretched by taking the humerus into extreme adduction behind the back. Extreme horizontal adduction stretches the posterior deltoid.

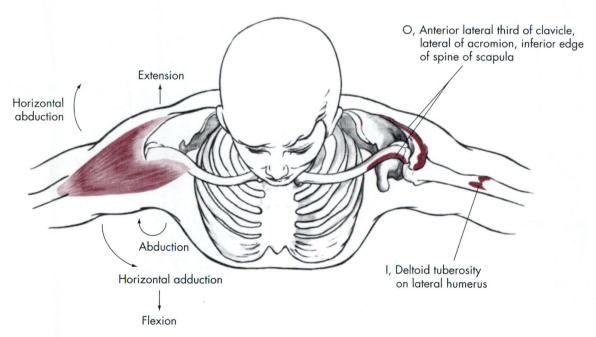

FIG. 3.8 ● Deltoid muscle, superior view. *O*, Origin; *I*, insertion.

Coracobrachialis muscle   FIG. 3.9

(kor-a-ko-bra′ki-a′lis)

Origin

Coracoid process of the scapula

Insertion

Middle of the medial border of the humeral shaft

Action

Flexion of the glenohumeral joint
Adduction of the glenohumeral joint
Horizontal adduction of the glenohumeral joint

Palpation

Difficult to palpate

Innervation

Musculocutaneous nerve (C5–7)

Application, strengthening, and flexibility

The coracobrachialis is not a powerful muscle, but it does assist in flexion and adduction and is most functional in moving the arm horizontally toward and across the chest. It is best strengthened by horizontally adducting the arm against resistance, such as in bench pressing. It may also be strengthened by performing lat pulls, which are defined on page 54.

The coracobrachialis is best stretched in extreme horizontal abduction, although extreme extension also stretches this muscle.

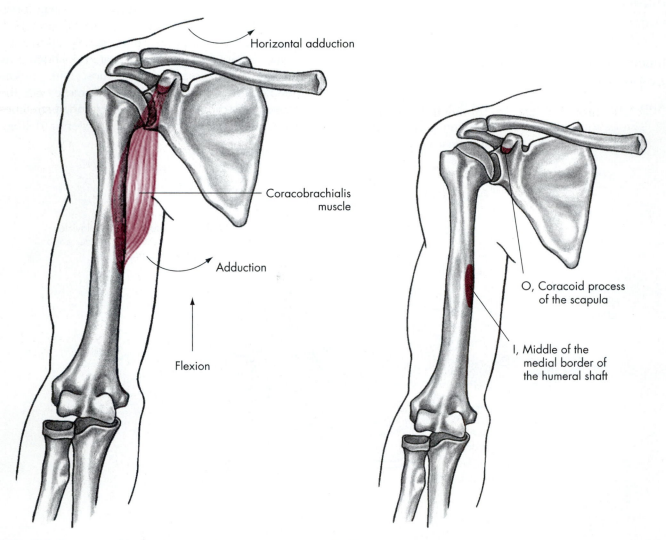

FIG. 3.9 ● Coracobrachialis muscle, anterior view. *O*, Origin; *I*, insertion.

Supraspinatus muscle   FIG. 3.10

(su´pra-spi-na´tus)

Origin

Medial two-thirds of the supraspinatus fossa

Insertion

Superiorly on the greater tubercle of the humerus

Action

Weak abduction and stabilization of the humeral
  head in the glenoid fossa

Palpation

Cannot be palpated; lies under the deltoid muscle
  distally and under the trapezius proximally

Innervation

Suprascapular nerve (C5)

Application, strengthening, and flexibility

The supraspinatus muscle holds the head of the
humerus in the glenoid fossa. In throwing movements, it provides important dynamic stability by
maintaining the proper relationship between the
humeral head and the glenoid fossa. In the cocking phase of throwing, there is a tendency for the
humeral head to subluxate anteriorly. In the
follow-through phase, the humeral head tends to
move posteriorly.

The supraspinatus, along with the other rotator
cuff muscles, must have excellent strength and
endurance to prevent abnormal and excessive
movement of the humeral head in the fossa.

The supraspinatus is the most often injured rotator cuff muscle. Acute severe injuries may occur
with trauma to the shoulder. However, mild to
moderate strains or tears often occur with athletic
activity, particularly if the activity involves repetitious overhead movements, such as throwing or
swimming.

Injury or weakness in the supraspinatus may be
detected when the athlete attempts to substitute
the scapula elevators and upward rotators to obtain humeral abduction. An inability to smoothly
abduct the arm against resistance is indicative of
possible rotator cuff injury.

The supraspinatus muscle may be called into
play whenever the middle fibers of the deltoid
muscle are used. An "empty-can exercise" may be
used to emphasize supraspinatus action. This is
performed by internally rotating the humerus, followed by abducting the arm to 90 degrees in a 30-
to 45-degree horizontally adducted position, as if
one were emptying a can.

Adducting the arm behind the back with the
shoulder internally rotated and extended stretches
the supraspinatus.

FIG. 3.10 ● Supraspinatus muscle, posterior
view. *O*, Origin; *I*, insertion.

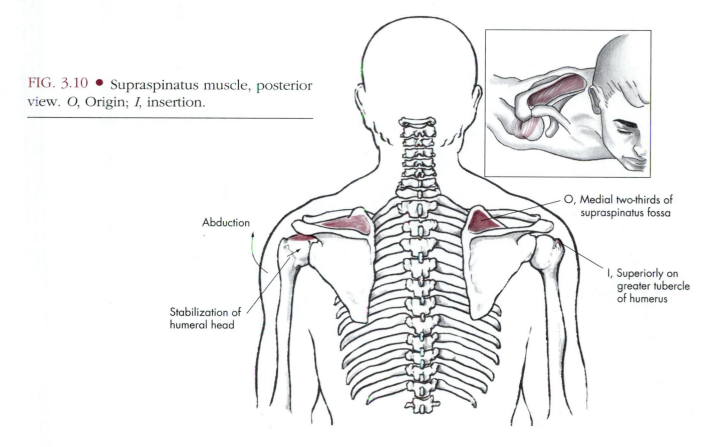

Abduction

Stabilization of
humeral head

O, Medial two-thirds of
supraspinatus fossa

I, Superiorly on
greater tubercle
of humerus

Infraspinatus muscle   FIG. 3.11

(in´fra-spi-na´tus)

Origin

Medial aspect of the infraspinatus fossa just below the spine of the scapula

Insertion

Posteriorly on the greater tubercle of the humerus

Action

External rotation of the glenohumeral joint
Horizontal abduction of the glenohumeral joint
Extension of the glenohumeral joint
Stabilization of the humeral head in the glenoid fossa

Palpation

Immediately below the spine of the scapula and the posterior fibers of the deltoid muscle

Innervation

Suprascapular nerve (C5, 6)

Application, strengthening, and flexibility

The infraspinatus and teres minor muscles are effective when the rhomboid muscles stabilize the scapula. When the humerus is rotated outward, the rhomboid muscles flatten the scapula to the back and fixate it so that the humerus may be rotated.

The infraspinatus is vital to maintaining the posterior stability of the glenohumeral joint. It is the most powerful of the external rotators and is the second most commonly injured rotator cuff muscle.

Exercises in which the arms are pulled down bring the infraspinatus, teres major, and latissimus dorsi into powerful contraction. Chinning, rope climbing, and dips on parallel bars are good exercises for these muscles. Both the infraspinatus and the teres minor can best be strengthened by externally rotating the arm against resistance in the 0-degree abducted position and the 90-degree abducted position.

Stretching of the infraspinatus is accomplished with internal rotation and extreme horizontal adduction.

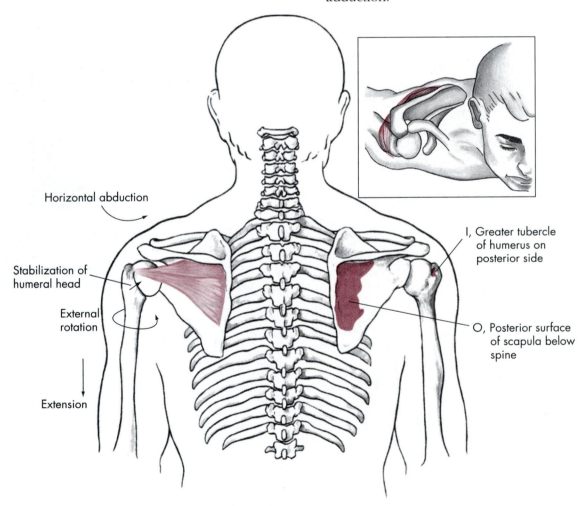

Horizontal abduction

Stabilization of humeral head

External rotation

Extension

I, Greater tubercle of humerus on posterior side

O, Posterior surface of scapula below spine

FIG. 3.11 • Infraspinatus muscle, posterior view. *O*, Origin; *I*, insertion.

Teres minor muscle   FIG. 3.12

(te´rez mi´nor)

Origin

Posteriorly on the upper and middle aspect of the
lateral border of the scapula

Insertion

Posteriorly on the greater tubercle of the humerus

Action

External rotation of the glenohumeral joint
Horizontal abduction of the glenohumeral joint
Extension of the glenohumeral joint
Stabilization of the humeral head in the glenoid
fossa

Palpation

Between the posterior deltoid and the lateral
scapula border

Innervation

Axillary nerve (C5, 6)

Application, strengthening, and flexibility

The teres minor functions very similarly to the infraspinatus in providing dynamic posterior stability to the glenohumeral joint. Both of these muscles perform the same actions together. The teres
minor is strengthened with the same exercises
that are used in strengthening the infraspinatus.

The teres minor is stretched similarly to the infraspinatus by internally rotating the shoulder
while moving into extreme horizontal adduction.

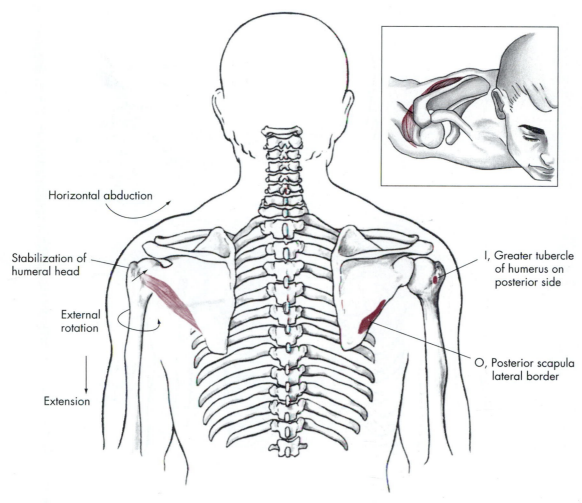

FIG. 3.12 ● Teres minor muscle, posterior view. *O*, Origin; *I*, insertion.

Subscapularis muscle   FIG. 3.13

(sub-skap-u-la´ris)

Origin

Entire anterior surface of the subscapular fossa

Insertion

Lesser tubercle of the humerus

Action

Internal rotation of the glenohumeral joint
Adduction of the glenohumeral joint
Extension of the glenohumeral joint
Stabilization of the humeral head in the glenoid
  fossa

Palpation

Cannot be palpated

Innervation

Upper and lower subscapular nerve (C5, 6)

Application, strengthening, and flexibility

The subscapularis muscle, another rotator cuff muscle, holds the head of the humerus in the glenoid fossa from in front and below. It acts with the latissimus dorsi and teres major muscles in its typical movement but is less powerful in its action because of its proximity to the joint. The muscle also requires the help of the rhomboid in stabilizing the scapula to make it effective in the movements described. The subscapularis is relatively hidden behind the rib cage in its location on the anterior aspect of the scapula in the subscapular fossa. It may be strengthened with exercises similar to those used for the latissimus dorsi and teres major, such as rope climbing and lat pulls. A specific exercise for its development is done by internally rotating the arm against resistance in the beside-the-body position at 0 degrees of glenohumeral abduction.

External rotation with the arm adducted by the side stretches the subscapularis.

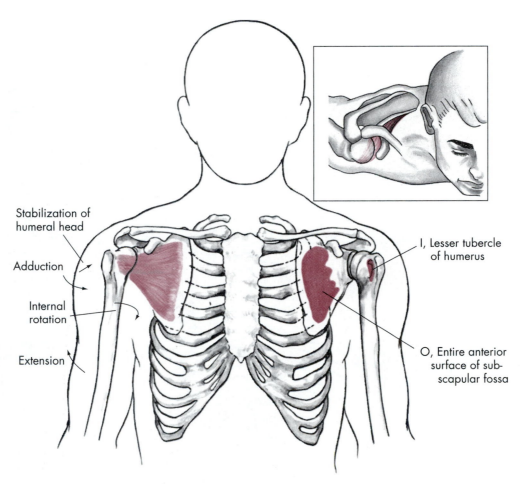

Stabilization of
humeral head

Adduction

Internal
rotation

Extension

I, Lesser tubercle
of humerus

O, Entire anterior
surface of sub-
scapular fossa

FIG. 3.13 ● Subscapularis muscle, anterior view. *O*, Origin; *I*, insertion.

Teres major muscle   FIG. 3.14

(te´rez ma´jor)

Origin

Posteriorly on the inferior third of the lateral border of the scapula and just superior to the inferior angle

Insertion

Medial lip of the intertubercular groove of the humerus

Action

Extension of the glenohumeral joint, particularly from the flexed position to the posteriorly extended position

Internal rotation of the glenohumeral joint

Adduction of the glenohumeral joint, particularly from the abducted position down to the side and toward the midline of the body

Palpation

Posterior scapular surface, moving diagonally upward from the inferior angle of the scapula

Innervation

Lower subscapular nerve (C5, 6)

Application, strengthening, and flexibility

The teres major muscle is effective only when the rhomboid muscles stabilize the scapula or move the scapula in downward rotation. Otherwise, the scapula would move forward to meet the arm.

This muscle works effectively with the latissimus dorsi. It assists the latissimus dorsi, pectoralis major, and subscapularis in adducting, internally rotating, and extending the humerus. It is said to be the latissimus dorsi's "little helper." It may be strengthened by lat pulls, rope climbing, and internal rotation exercises against resistance.

Externally rotating the shoulder in a 90-degree abducted position stretches the teres major.

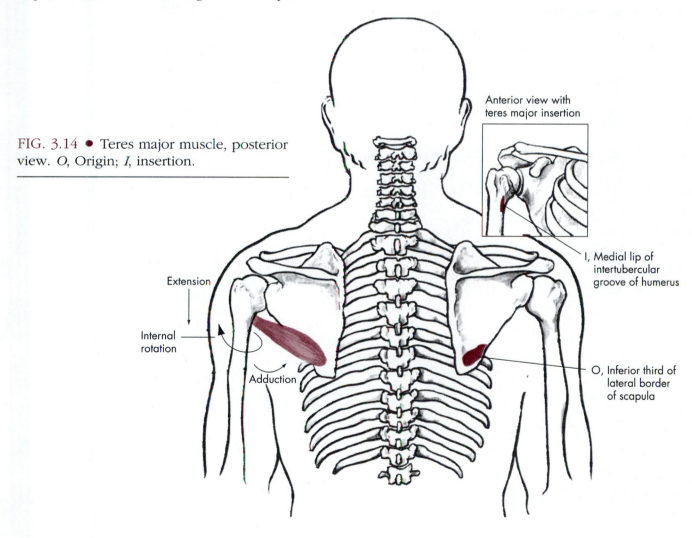

FIG. 3.14 ● Teres major muscle, posterior view. *O*, Origin; *I*, insertion.

Anterior view with teres major insertion

I, Medial lip of intertubercular groove of humerus

Extension

Internal rotation

Adduction

O, Inferior third of lateral border of scapula

Latissimus dorsi muscle   FIG. 3.15

(lat-is´i-mus dor´si)

Origin

Posterior crest of the ilium, back of the sacrum and
  spinous processes of the lumbar and lower six
  thoracic vertebrae (T6–12); slips from the lower
  three ribs

Insertion

Medial side of the intertubercular groove of the
  humerus

Action

Adduction of the glenohumeral joint
Extension of the glenohumeral joint
Internal rotation of the glenohumeral joint
Horizontal abduction of the glenohumeral joint

Palpation

Posterolateral aspect of the trunk below the armpit

Innervation

Thoracodorsal nerve (C6–8)

Application, strengthening, and flexibility

The latissimus dorsi muscle has strong action in
adduction of the humerus. Due to the upward
rotation of the scapula that accompanies glenohumeral abduction, the latissimus effectively
downwardly rotates the scapula by way of its action in pulling the entire shoulder girdle downward in active glenohumeral adduction. It is one
of the most important extensor muscles of the
humerus and contracts powerfully in chinning.

Exercises in which the arms are pulled down
bring the latissimus dorsi muscle into powerful contraction. Chinning, rope climbing, dips on
parallel bars, and other uprise movements on the
horizontal bar are good examples. In barbell exercises, the basic rowing and pullover exercises
are good for developing the "lats." Pulling the bar
of an overhead pulley system down toward the
shoulders, known as "lat pulls," is a common exercise for this muscle.

The latissimus dorsi is stretched with the teres
major when the shoulder is externally rotated
while in a 90-degree abducted position. This
stretch may be accentuated further by abducting
the shoulder fully while maintaining external rotation and then laterally flexing and rotating the
trunk to the opposite side.

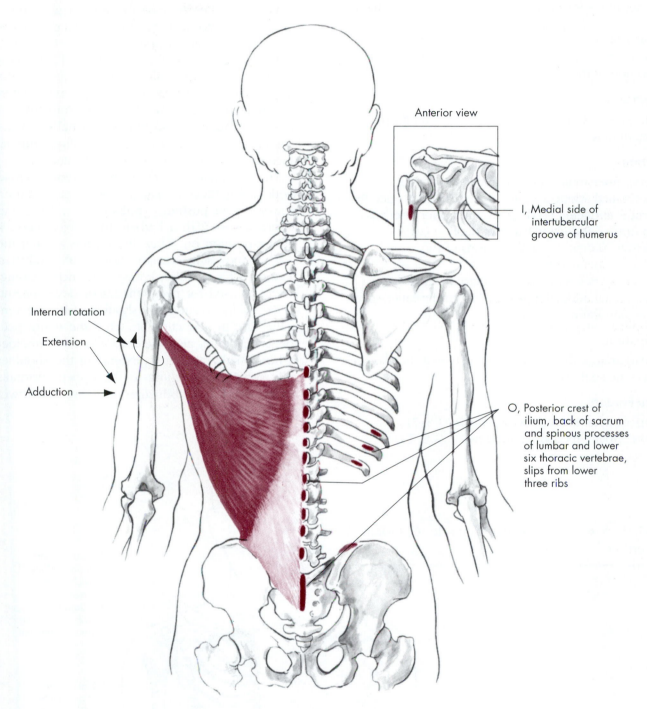

Anterior view

I, Medial side of
intertubercular
groove of humerus

Internal rotation

Extension

Adduction

O, Posterior crest of
ilium, back of sacrum
and spinous processes
of lumbar and lower
six thoracic vertebrae,
slips from lower
three ribs

FIG. 3.15 • Latissimus dorsi muscle, *O*, Origin; *I*, insertion.

Pectoralis major muscle   FIG. 3.16

(pek-to-ra´lis ma´jor)

Origin

Upper fibers (clavicular head): medial half of the anterior surface of the clavicle

Lower fibers (sternal head): anterior surfaces of the costal cartilage of the first six ribs, and adjacent portion of the sternum

Insertion

Flat tendon 2 or 3 inches wide to the outer lip of the intertubercular groove of the humerus

Action

Upper fibers (clavicular head): internal rotation, horizontal adduction, flexion, abduction (once the arm is abducted 90 degrees, the upper fibers assist in further abduction), and adduction (with the arm below 90 degrees of abduction) of the glenohumeral joint

Lower fibers (sternal head): internal rotation, horizontal adduction, extension, and adduction of the glenohumeral joint

Palpation

Broad area of the chest region between the clavicle and the sixth rib

Innervation

Upper fibers: lateral pectoral nerve (C5–7)
Lower fibers: medial pectoral nerve (C8, T1)

Application, strengthening, and flexibility

The pectoralis major muscle aids the serratus anterior muscle in drawing the scapula forward as it moves the humerus in flexion and internal rotation. Even though the pectoralis major is not attached to the scapula, it is effective in this scapula protraction because of its anterior pull on the humerus, which joins to the scapula at the glenohumeral joint. Typical action is shown in throwing a baseball. As the glenohumeral joint is flexed, the humerus is internally rotated and the scapula is drawn forward with upward rotation. It also works as a helper of the latissimus dorsi muscle when extending and adducting the humerus from a raised position.

The pectoralis major and the anterior deltoid work closely together. The pectoralis major is used powerfully in push-ups, pull-ups, throwing, and tennis serves. With a barbell, the subject takes a supine position on a bench with the arms at the side and moves the arms to a horizontally adducted position. This exercise, known as bench pressing, is widely used for pectoralis major development.

Externally rotating the shoulder with the arm at the side in adduction stretches the entire pectoralis major. It is also stretched when the shoulder is horizontally abducted. Extending the shoulder fully provides stretching to the upper pectoralis major, while full abduction stretches the lower pectoralis major.

FIG. 3.16 ● Pectoralis major muscle, *O*, Origin; *I*, insertion.

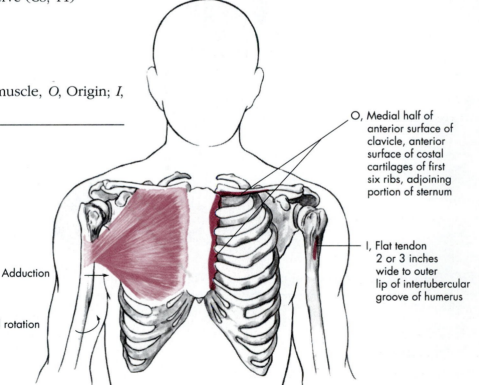

Adduction

Internal rotation

O, Medial half of anterior surface of clavicle, anterior surface of costal cartilages of first six ribs, adjoining portion of sternum

I, Flat tendon 2 or 3 inches wide to outer lip of intertubercular groove of humerus

Web sites

Anatomy & Physiology Tutorials:

www.gwc.maricopa.edu/class/bio201/index.htm

Electronic Textbook of Hand Surgery:

http://www.e-hand.com/default.htm

Slides and illustrations of upper extremity musculoskeletal anatomy

Radiologic Anatomy Browser:

radlinux1.usuf1.usuhs.mil/rad/iong/index.html

This site has numerous radiological views of the musculoskeletal system.

University of Arkansas Medical School Gross Anatomy for Medical Students:

anatomy.uams.edu/htmlpages/anatomyhtml/gross.html

Dissections, anatomy tables, atlas images, links, etc.

Loyola University Medical Center: Structure of the Human Body:

www.meddean.luc.edu/lumen/MedEd/ GrossAnatomy/GA.html

An excellent site with many slides, dissections, tutorials, etc., for the study of human anatomy.

Wheelless' Textbook of Orthopaedics:

www.medmedia.com/

This site has an extensive index of links to the fractures, joints, muscles, nerves, trauma, medications, medical topics, and lab tests, as well as links to orthopedic journals and other orthopedic and medical news.

Premiere Medical Search Engine:

www.medsite.com

This site allows the reader to enter any medical condition and it will search the net to find relevant articles.

Arthroscopy.Com.:

www.arthroscopy.com/sports.htm

Patient information on various musculoskeletal problems of the upper and lower extremity

Anatomy of the Shoulder Tutorial:

www.ncl.ac.uk~nccc/tutorials/shoulder/

Virtual Hospital:

www.vh.org

Numerous slides, patient information, etc.

Dynamic Human version 2.0 CD-ROM: The Visual Guide to Anatomy & Physiology:

www.mhhe.com/biosci/ap/dynamichuman2/

Web site that accompanies this CD-ROM

Dynamic Human CD activities

1. Review anatomical landmarks as well as origins/insertions for the clavicle and scapula by clicking on **skeletal**; then **anatomy**; then **gross anatomy, pectoral girdle**, and **clavicle**; and then **scapula**.
2. Review anatomical landmarks as well as origins/insertions for the humerus, radius, and ulna as they pertain to the shoulder joint by clicking on **skeletal**; then **anatomy**; then **gross anatomy, upper limbs**, and **humerus**; and then **radius/ulna**.
3. Review each of the muscles from this chapter by clicking on **muscular, anatomy**, and **body regions** and then **pectoral girdle** and **upper arms**.
4. What effect do you think a dislocated shoulder has on the muscles and ligaments on the shoulder? To access the dislocated shoulder, click on **skeletal**, then **clinical concepts**, and then **dislocated shoulder**.
5. Review each of the muscles from this chapter. Then click on **muscular**, then **clinical concepts**, and then **MRI of the rotator cuff**. What are the clinical implications of the pathology in the second image? What is the effect on the glenohumeral joint?

Worksheet exercises

As an aid to learning, for in-class or out-of-class assignments, or for testing, tear-out worksheets are found at the end of the text (pp. 254 and 255).

Skeletal worksheet (no. 1)

Draw and label on the worksheet the following muscles:
a. Deltoid
b. Supraspinatus
c. Subscapularis
d. Teres major
e. Infraspinatus
f. Teres minor
g. Latissimus dorsi
h. Pectoralis major
i. Coracobrachialis

Human figure worksheet (no. 2)

Label and indicate with arrows the following movements of the shoulder joint:

a. Abduction
b. Adduction
c. Flexion
d. Extension
e. Horizontal adduction
f. Horizontal abduction
g. Internal rotation
h. External rotation

Laboratory and review exercises

1. Locate the following parts of the humerus and scapula on a human skeleton and on a subject:
   a. Greater tubercle
   b. Lesser tubercle
   c. Neck
   d. Shaft
   e. Intertubercular groove
   f. Medial epicondyle
   g. Lateral epicondyle
   h. Trochlea
   i. Capitulum
   j. Supraspinatus fossa
   k. Infraspinatus fossa
   l. Spine of the scapula

2. How and where can the following muscles be palpated on a human subject?
   a. Deltoid
   b. Teres major
   c. Infraspinatus
   d. Teres minor
   e. Latissimus dorsi
   f. Pectoralis major (upper and lower)
      NOTE: Using the pectoralis major muscle, indicate how various actions allow muscle palpation.

3. Demonstrate and locate on a human subject the muscles that are primarily used in the following shoulder joint movements:
   a. Abduction
   b. Adduction
   c. Flexion
   d. Extension
   e. Horizontal adduction
   f. Horizontal abduction
   g. External rotation
   h. Internal rotation

4. List the planes in which each of the following glenohumeral joint movements occur. List the respective axis of rotation for each movement in each plane.
   a. Abduction
   b. Adduction
   c. Flexion
   d. Extension
   e. Horizontal adduction
   f. Horizontal abduction
   g. External rotation
   h. Internal rotation

5. Why is it essential that both anterior and posterior muscles of the shoulder joint be properly developed? What are some activities or sports that would cause unequal development? equal development?

6. Using an articulated skeleton, compare the relationship of the greater tubercle to the undersurface of the acromion in each of the following situations:
   a. Flexion with the humerus internally rotated versus externally rotated
   b. Abduction with the humerus internally versus externally rotated
   c. Horizontal adduction with the humerus internally versus externally rotated

7. What practical application do the activities or sports in question #5 support if
   a. The rotator cuff muscles are not functioning properly due to fatigue or lack of appropriate strength and endurance?
   b. The scapula stabilizers are not functioning properly due to fatigue or lack of strength and endurance?

8. Pair up with a partner with the back exposed. Use your hand to grasp your partner's right scapula along the lateral border to prevent scapula movement. Have your partner slowly abduct the glenohumeral joint as much as possible. Note the difference in total abduction possible normally versus when you restrict movement of the scapula. Repeat the same exercise, except hold the inferior angle of the scapula tightly against the chest wall while you have your partner internally rotate the humerus. Note the difference in total internal rotation possible normally versus when you restrict movement of the scapula.

9. Analyze movements and muscles in both the shoulder girdle and joint when the following activities are performed:
   a. Chinning (actual pull)
   b. Throwing a baseball (throw only)

c. Batting a baseball (striking ball)

d. Performing a push-up (actual push)

10. Fill in the antagonistic muscle action chart by listing the muscle(s) or parts of muscles that are antagonist in their actions to the muscles in the left column.

11. Fill in the muscle analysis chart by listing the muscles primarily involved in each joint movement.

Antagonistic muscle action chart • Glenohumeral joint

Agonist	Antagonist
Deltoid (anterior fibers)	
Deltoid (middle fibers)	
Deltoid (posterior fibers)	
Supraspinatus	
Subscapularis	
Teres major	
Infraspinatus/teres minor	
Latissimus dorsi	
Pectoralis major (upper fibers)	
Pectoralis major (lower fibers)	
Coracobrachialis	

Muscle analysis chart • Shoulder girdle and shoulder joint

Shoulder Girdle	Shoulder Joint
Adduction	Extension
Abduction	Flexion
Elevation	Horizontal adduction
Depression	Horizontal abduction
Upward rotation	Abduction
Downward rotation	Adduction
	External rotation
	Internal rotation
	Diagonal abduction (overhand activities)
	Diagonal adduction (overhand activities)

References

Andrews JR, Wilk KE: *The athlete's shoulder,* New York, 1994, Churchill Livingstone.

Andrews JR, Zarins B, Wilk KE: *Injuries in baseball,* Philadelphia, 1988, Lippincott-Raven.

Garth WP, et al: Occult anterior subluxations of the shoulder in noncontact sports, *American Journal of Sports Medicine* 15:579, November-December 1987.

Hislop HJ, Montgomery J: *Daniels and Worthingham's muscle testing: techniques of manual examination,* ed 6, Philadelphia, 1995, Saunders.

Perry JF, Rohe DA, Garcia AO: *The kinesiology workbook,* Philadelphia, 1992, Davis.

Rasch PJ: *Kinesiology and applied anatomy,* ed 7, Philadelphia, 1989, Lea & Febiger.

Sieg KW, Adams SP: *Illustrated essentials of musculoskeletal anatomy,* ed 2, Gainesville, FL, 1985, Megabooks.

Smith LK, Weiss EL, Lehmkuhl LD: *Brunnstrom's clinical kinesiology,* ed 5, Philadelphia, 1996, Davis.

Stacey E: Pitching injuries to the shoulder, *Athletic Journal* 65:44, January 1984.

The elbow and radioulnar joints

4

Objectives

- To identify on a human skeleton selected bony features of the elbow and radioulnar joints

- To label selected bony features on a skeletal chart

- To draw and label the muscles on a skeletal chart

- To palpate the muscles on a human subject and list their antagonists

- To list the planes of motion and their respective axes of rotation

- To organize and list the muscles that produce the primary movements of the elbow joint and the radioulnar joint

Almost any movement of the upper extremity will involve the elbow and radioulnar joints. Quite often, these joints are grouped together because of their close anatomical relationship. For this reason, novice students may confuse motions of the elbow with those of the radioulnar joint. In addition, radioulnar joint motion may be incorrectly attributed to the wrist joint because it appears to occur there. However, with close inspection, the elbow joint and its movements can be clearly distinguished from those of the radioulnar joints, just as the radioulnar movements can be distinguished from those of the wrist.

Bones

The ulna is much larger proximally than the radius (Fig. 4.1), but distally the radius is much larger than the ulna (see Fig. 5.1 in Chapter 5). The scapula and humerus serve as the proximal attachments for the muscles that flex and extend the elbow. The ulna and radius serve as the distal attachments for the same muscles. The scapula, humerus, and ulna serve as proximal attachments for the muscles that pronate and supinate the radioulnar joints. The distal attachments of the radioulnar joint muscles are located on the radius.

Joints

The elbow joint is classified as a ginglymus or hinge-type joint that allows only flexion and exten-sion (Fig. 4.1). The elbow may actually be thought of as two interrelated joints, the humeroulnar and the radiohumeral joints. Elbow motions primarily involve movement between the articular surfaces of the humerus and ulna—specifically, the humeral trochlear fitting into the trochlear notch of the ulna. The head of the radius has a relatively small amount of contact with the capitulum of the humerus. As the elbow reaches full extension, the olecranon process of the ulna is received by the olecranon fossa of the humerus. This arrangement provides increased joint stability when the elbow is fully extended.

As the elbow flexes approximately 20 degrees or more, its bony stability is somewhat unlocked, allowing for more side-to-side laxity. The stability of the elbow in flexion is more dependent on the collateral ligaments, such as the lateral or radial collateral ligament and especially the medial or ulnar collateral ligament (Fig. 4.2). The ulnar collateral ligament is critical to providing medial support to prevent the elbow from abducting (not a normal movement of the elbow) when stressed in physical activity. Many contact sports, particularly sports with throwing activities, place stress on the medial aspect of the joint, resulting in injury. The radial collateral ligament on the opposite side provides lateral stability and is rarely injured. Additionally, the annular ligament is located

FIG. 4.1 ● Elbow joint. **A**, anterior view; **B**, lateral view; **C**, medial view.

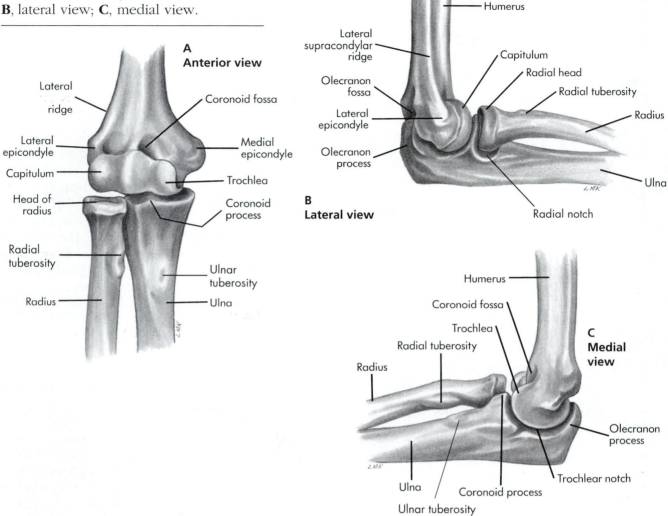

A Anterior view

Lateral ridge

Coronoid fossa

Lateral epicondyle

Medial epicondyle

Capitulum

Trochlea

Head of radius

Coronoid process

Radial tuberosity

Ulnar tuberosity

Radius

Ulna

Humerus

Lateral supracondylar ridge

Capitulum

Radial head

Olecranon fossa

Radial tuberosity

Lateral epicondyle

Radius

Olecranon process

Ulna

B Lateral view

Radial notch

Humerus

Coronoid fossa

Trochlea

Radial tuberosity

Radius

C Medial view

Olecranon process

Trochlear notch

Ulna

Coronoid process

Ulnar tuberosity

FIG. 4.2 ● Elbow joint with ligaments detailed. **A**, Lateral view; **B**, medial view.

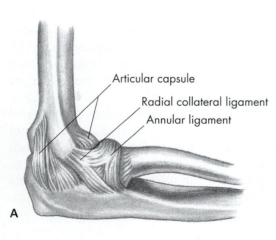

Articular capsule

Radial collateral ligament

Annular ligament

A

Lateral view

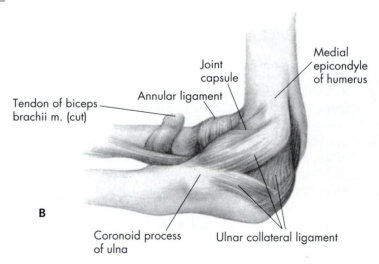

Joint capsule

Medial epicondyle of humerus

Annular ligament

Tendon of biceps brachii m. (cut)

B

Coronoid process of ulna

Ulnar collateral ligament

Medial view

laterally, providing a sling effect around the radial head to secure its stability.

The elbow is capable of moving from 0 degrees of extension to approximately 145 to 150 degrees of flexion.

The radioulnar joint is classified as a trochoid or pivot-type joint. The radial head rotates around in its location at the proximal ulna. This rotary movement is accompanied by the distal radius rotating around the distal ulna. The radial head is maintained in its joint by the annular ligament. The radioulnar joint can supinate approximately 80 to 90 degrees from the neutral position. Pronation varies from 70 to 90 degrees.

Movements   FIG. 4.3

Elbow movements

Flexion: movement of the forearm to the shoulder by bending the elbow to decrease its angle

Extension: movement of the forearm away from the shoulder by straightening the elbow to increase its angle

Radioulnar joint movements

Pronation: internal rotary movement of the radius on the ulna that results in the hand moving from the palm-up to the palm-down position

Supination: external rotary movement of the radius on the ulna that results in the hand moving from the palm-down to the palm-up position

FIG. 4.3 ● Movements of the elbow and radioulnar joint. **A**, Elbow flexion; **B**, elbow extension; **C**, radioulnar pronation; **D**, radioulnar supination.

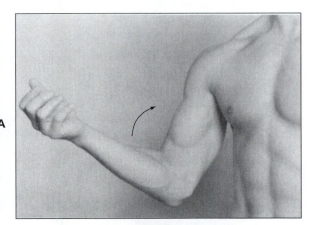

A

Flexion

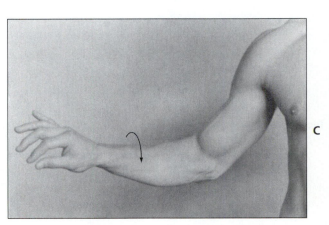

C

Pronation

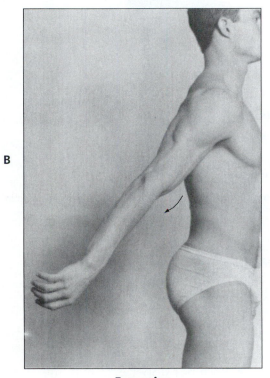

B

Extension

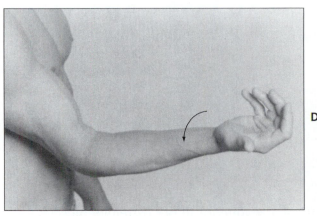

D

Supination

Muscles

The muscles of the elbow and radioulnar joints may be more clearly understood when separated by function. The elbow flexors, located anteriorly, are the biceps brachii, the brachialis, and the brachioradialis, with some weak assistance from the pronator teres (Fig. 4.4).

The triceps brachii, located posteriorly, is the primary elbow extensor, with assistance provided by the anconeus (Fig. 4.5). The pronator group, located anteriorly, consists of the pronator teres, the pronator quadratus, and the brachioradialis. The brachioradialis also assists with supination, which is controlled mainly by the supinator muscle and the biceps brachii. The supinator muscle is located posteriorly.

A common problem associated with the muscles of the elbow is "tennis elbow," which usually involves the extensor digitorum muscle near its origin on the lateral epicondyle. This condition, known technically as lateral epicondylitis, is quite frequently associated with gripping and lifting activities. Medial epicondylitis, a somewhat less common problem frequently referred to as "golfer's elbow," is associated with the wrist flexor and pronator group near their origin on the medial epicondyle. Both of these conditions involve muscles which cross the elbow but act primarily on the wrist and hand. These muscles will be addressed in Chapter 5.

Elbow and radioulnar joint muscles—location

Anterior
    Primarily flexion and pronation
        Biceps brachii
        Brachialis
        Brachioradialis
        Pronator teres
        Pronator quadratus
Posterior
    Primarily extension and supination
        Triceps brachii
        Anconeus
        Supinator

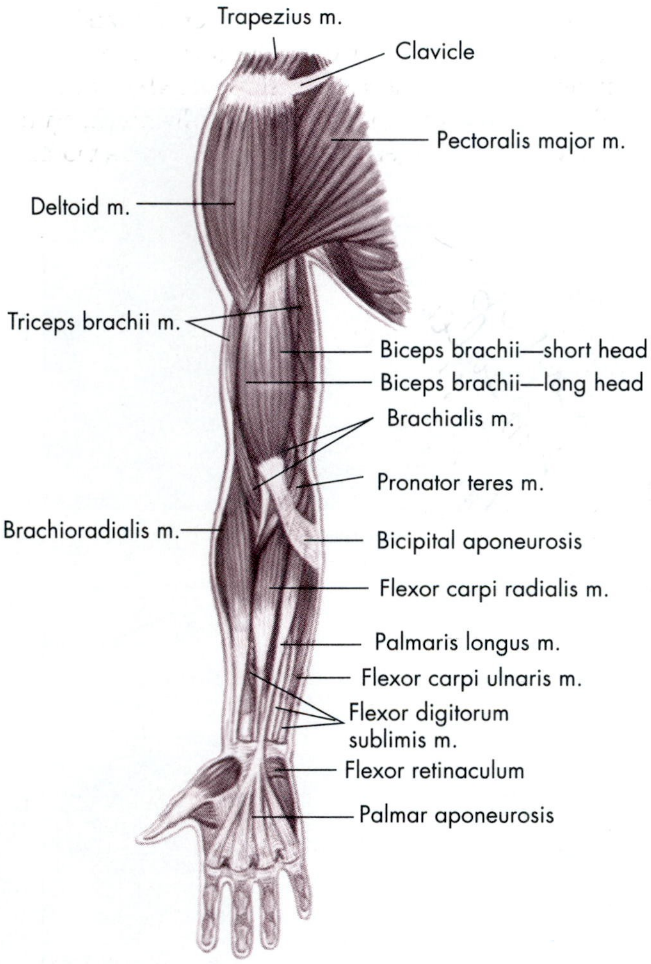

Trapezius m.
Clavicle
Pectoralis major m.
Deltoid m.
Triceps brachii m.
Biceps brachii—short head
Biceps brachii—long head
Brachialis m.
Pronator teres m.
Brachioradialis m.
Bicipital aponeurosis
Flexor carpi radialis m.
Palmaris longus m.
Flexor carpi ulnaris m.
Flexor digitorum sublimis m.
Flexor retinaculum
Palmar aponeurosis

FIG. 4.4 ● Anterior upper extremity muscles.

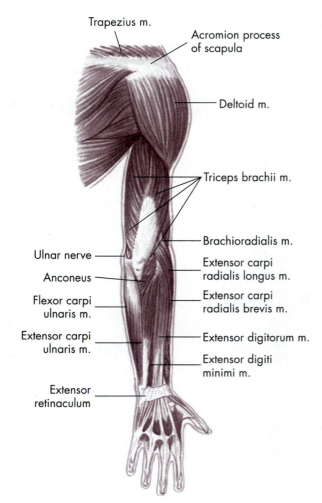

Trapezius m.
Acromion process of scapula
Deltoid m.
Triceps brachii m.
Ulnar nerve
Anconeus
Flexor carpi ulnaris m.
Extensor carpi ulnaris m.
Extensor retinaculum
Brachioradialis m.
Extensor carpi radialis longus m.
Extensor carpi radialis brevis m.
Extensor digitorum m.
Extensor digiti minimi m.

FIG. 4.5 ● Posterior upper extremity muscles.

Biceps brachii muscle   FIG. 4.6

(bi´seps bra´ki-i)

Origin

Two heads

Long head: supraglenoid tubercle above the
  superior lip of the glenoid fossa

Short head: coracoid process of the scapula and
  upper lip of the glenoid fossa

Insertion

Tuberosity of the radius and bicipital aponeurosis

Action

Flexion of the elbow
Supination of the forearm
Weak flexion of the shoulder joint

Palpation

Easily palpated on the anterior aspect of the
  humerus and elbow

Innervation

Musculocutaneous nerve (C5, 6)

Application, strengthening, and flexibility

The biceps is commonly known as a two-joint
(shoulder and elbow), or biarticular, muscle. However, technically it should be considered a three-
joint (multiarticular) muscle—shoulder, elbow, and
radioulnar. It is weak in actions of the shoulder
joint, although it does assist in providing dynamic
anterior stability to maintain the humeral head in
the glenoid fossa. It is more powerful in flexing
the elbow when the radioulnar joint is supinated.
It is also a strong supinator, particularly if the
elbow is flexed. Palms away from the face (pronation) decrease the effectiveness of the biceps,
partly as a result of the disadvantageous pull of the
muscle as the radius rotates. The same muscles are
used in elbow joint flexion, regardless of forearm
pronation or supination.

Flexion of the forearm with a barbell in the
hands, known as "curling," is an excellent exercise to develop the biceps brachii. This movement
can be performed one arm at a time with dumbbells or both arms simultaneously with a barbell.
Other activities in which there is powerful flexion
of the forearm are chinning and rope climbing.

Due to the multiarticular orientation of the biceps, all three joints must be positioned appropriately to achieve optimal stretching. The elbow
must be extended maximally with the shoulder in
full extension. The biceps may also be stretched
by beginning with full elbow extension and pro-
gressing into full horizontal abduction at approximately 70 to 110 degrees of shoulder abduction. In
all cases, the forearm should be fully pronated to
achieve maximal lengthening of the biceps brachii.

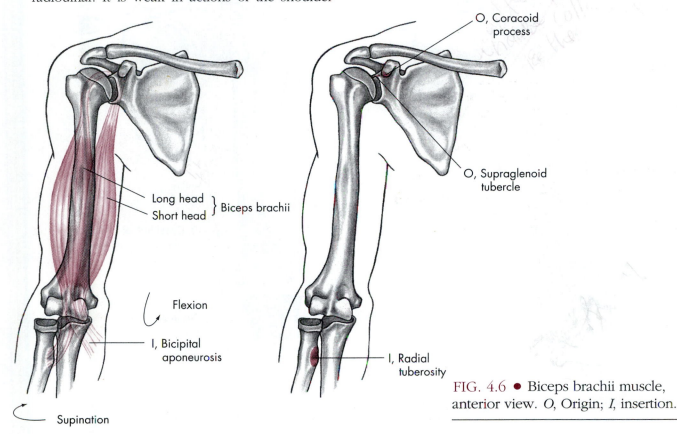

FIG. 4.6 ● Biceps brachii muscle,
anterior view. *O*, Origin; *I*, insertion.

Brachialis muscle   FIG. 4.7

(braˊki-aˊlis)

Origin
Distal half of the anterior portion of the humerus

Insertion
Coronoid process of the ulna

Action
True flexion of the elbow

Palpation
Lateral side of the upper arm under the biceps
  brachii muscle

Innervation
Musculocutaneous nerve and sometimes branches
  from radial and median nerves (C5, 6)

Application, strengthening, and flexibility

The brachialis muscle is used along with other
flexor muscles, whether in pronation or supination. It pulls on the ulna, which does not rotate,
thus making this muscle the only pure flexor of
this joint.

The brachialis muscle is called into action
whenever the elbow flexes. It is exercised along
with elbow curling exercises, as described for the
biceps brachii, pronator teres, and brachioradialis
muscles. Elbow flexion activities that occur with
the forearm pronated isolate the brachialis to
some extent by reducing the effectiveness of the
biceps brachii. Since the brachialis is a pure flexor
of the elbow, it can be stretched maximally only
by extending the elbow with the shoulder relaxed
and flexed. Forearm positioning should not affect
the stretch on the brachialis unless the forearm
musculature itself limits elbow extension, in
which case the forearm is probably best positioned in neutral.

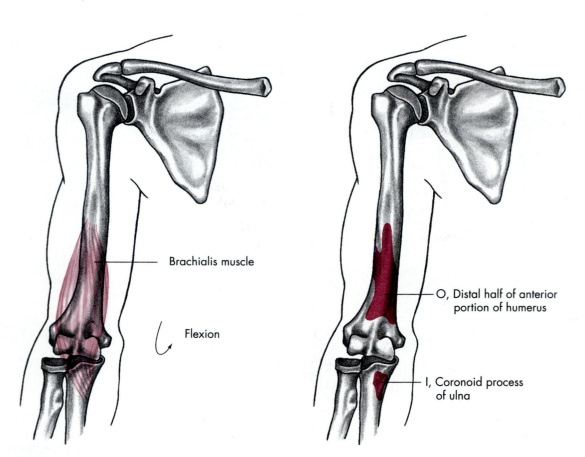

Brachialis muscle

Flexion

O, Distal half of anterior
portion of humerus

I, Coronoid process
of ulna

FIG. 4.7 ● Brachialis muscle, anterior view. *O*, Origin; *I*, insertion.

Brachioradialis muscle   FIG. 4.8

(braˊki-o-raˊdi-aˊlis)

Origin

Distal two-thirds of the lateral condyloid (supra-condylar) ridge of the humerus

Insertion

Lateral surface of the distal end of the radius at the styloid process

Action

Flexion of the elbow
Pronation from supinated position to neutral
Supination from pronated position to neutral

Palpation

On the lateral anterior side of the forearm

Innervation

Radial nerve (C5, 6)

Application, strengthening, and flexibility

The brachioradialis muscle acts as a flexor best in a midposition or neutral position between pronation and supination. In a supinated position of the forearm, it tends to pronate as it flexes. In a pronated position, it tends to supinate as it flexes. This muscle is favored in its action of flexion when the neutral position between pronation and supination is assumed, as previously suggested. Its insertion at the end of the radius makes it a strong elbow flexor. Its ability as a supinator decreases as the radioulnar joint moves toward neutral. Similarly, its ability to pronate decreases as the forearm reaches neutral.

The brachioradialis may be strengthened by performing elbow curls against resistance, particularly with the radioulnar joint in the neutral position. In addition, the brachioradialis may be developed by performing pronation and supination movements through the full range of motion against resistance.

The brachioradialis is stretched by maximally extending the elbow with the shoulder in flexion and the forearm in either maximal pronation or maximal supination.

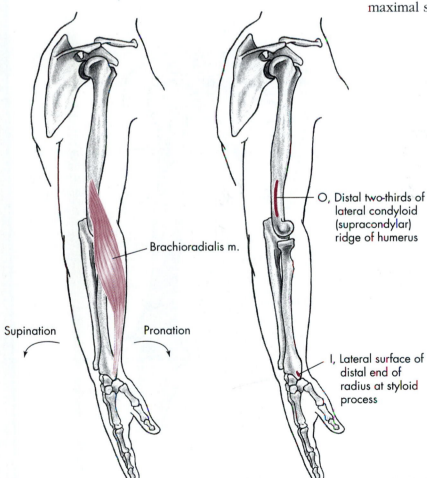

Brachioradialis m.

Supination   Pronation

O, Distal two-thirds of lateral condyloid (supracondylar) ridge of humerus

I, Lateral surface of distal end of radius at styloid process

FIG. 4.8 ● Brachioradialis muscle, lateral view. *O,* Origin; *I,* insertion.

Triceps brachii muscle FIG. 4.9

(tri´seps bra´ki-i)

Origin

Long head: infraglenoid tubercle below inferior lip of glenoid fossa of the scapula

Lateral head: upper half of the posterior surface of the humerus

Medial head: distal two-thirds of the posterior surface of the humerus

Insertion

Olecranon process of the ulna

Action

All heads: extension of the elbow

Long head: extension of the shoulder joint

Adduction of the shoulder joint

Palpation

Posterior and lateral aspects of the humerus

Innervation

Radial nerve (C7, 8)

Application, strengthening, and flexibility

Typical action of the triceps brachii is shown in push-ups when there is powerful extension of the elbow. It is used in hand balancing and in any pushing movement involving the upper extremity. The long head is an important extensor of the shoulder joint.

Two muscles extend the elbow—the triceps brachii and the anconeus. Push-ups demand strenuous contraction of these muscles. Dips on the parallel bars are more difficult to perform. Bench-pressing a barbell or a dumbbell is an excellent exercise. Overhead presses and triceps curls (elbow extensions from an overhead position) emphasize the triceps.

The triceps brachii should be stretched with both the shoulder and the elbow in maximal flexion.

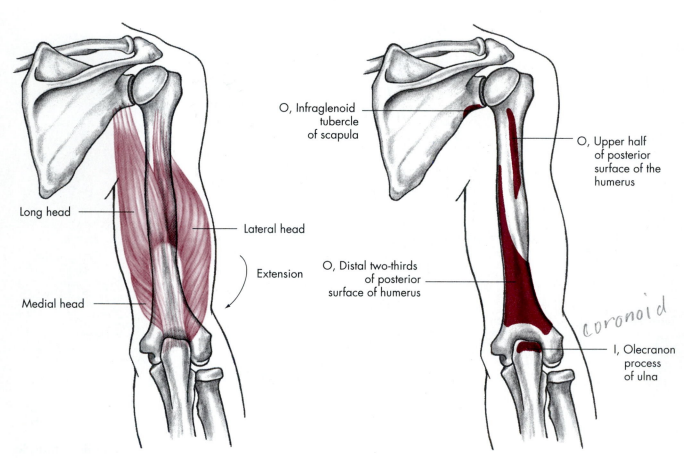

FIG. 4.9 ● Triceps brachii muscle, posterior view. *O,* Origin; *I,* insertion.

Anconeus muscle   FIG. 4.10

(an-ko´ne-us)

Origin

Posterior surface of the lateral condyle of the
humerus

Insertion

Posterior surface of the olecranon process of the
ulna

Action

Extension of the elbow

Palpation

Posterior lateral aspect of the olecranon process

Innervation

Radial nerve (C7, 8)

Application, strengthening, and flexibility

The chief function of the anconeus muscle is to
pull the synovial membrane of the elbow joint
out of the way of the advancing olecranon
process during extension of the elbow. It contracts along with the triceps brachii. It is strengthened with any elbow extension exercise against
resistance.

Maximal elbow flexion stretches the anconeus.

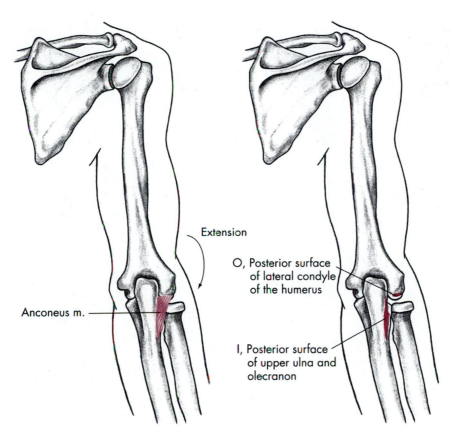

Extension

Anconeus m.

O, Posterior surface
of lateral condyle
of the humerus

I, Posterior surface
of upper ulna and
olecranon

FIG. 4.10 ● Anconeus muscle, posterior view. *O,* Origin; *I,* insertion.

Pronator teres muscle   FIG. 4.11

(pro-na´tor te´rez)

Origin

Distal part of the medial condyloid ridge of the humerus and medial side of the ulna

Insertion

Middle third of the lateral surface of the radius

Action

Pronation of the forearm
Weak flexion of the elbow

Palpation

Anteromedial surface of the proximal forearm

Innervation

Median nerve (C6, 7)

Application, strengthening, and flexibility

Typical movement of the pronator teres muscle is with the forearm pronating as the elbow flexes. Movement is weaker in flexion with supination. The use of the pronator teres alone in movement tends to bring the back of the hand to the face as it contracts. Pronation of the forearm with a dumbbell in the hand localizes action and develops the pronator teres muscle. The hammer exercise used for the supinator muscle may be modified to develop the pronator teres. In the beginning, the forearm is supported and the hand is free off the table edge. The hammer is again held suspended out of the ulnar side of the hand hanging toward the floor. The forearm is then pronated to the palm-down position to strengthen this muscle.

The elbow must be fully extended while taking the forearm into full supination to stretch the pronator teres.

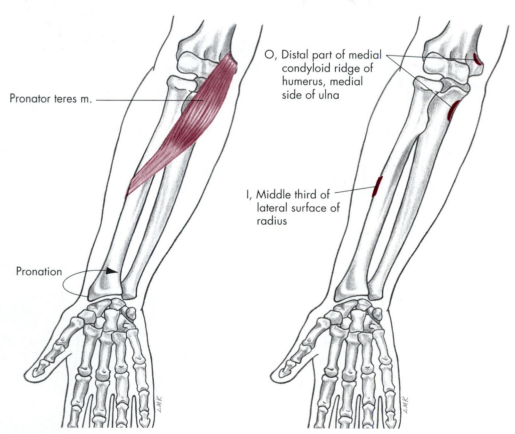

FIG. 4.11 ● Pronator teres muscle, anterior view. *O*, Origin; *I*, insertion.

Pronator quadratus muscle   FIG. 4.12

(pro-na´tor kwad-ra´tus)

Origin

Distal fourth of the anterior side of the ulna

Insertion

Distal fourth of the anterior side of the radius

Action

Pronation of the forearm

Palpation

Cannot be palpated

Innervation

Median nerve, (palmar interosseous branch) (C6, 7)

Application, strengthening, and flexibility

The pronator quadratus muscle works in pronating the forearm in combination with the triceps in extending the elbow. It is commonly used in turning a screwdriver, as in taking out a screw, when extension and pronation are needed. It is used also in throwing a screwball, when extension and pronation are needed. It may be developed with similar pronation exercises against resistance, as described for the pronator teres. The pronator quadratus is best stretched by using a partner to grasp the wrist and passively take the forearm into extreme supination.

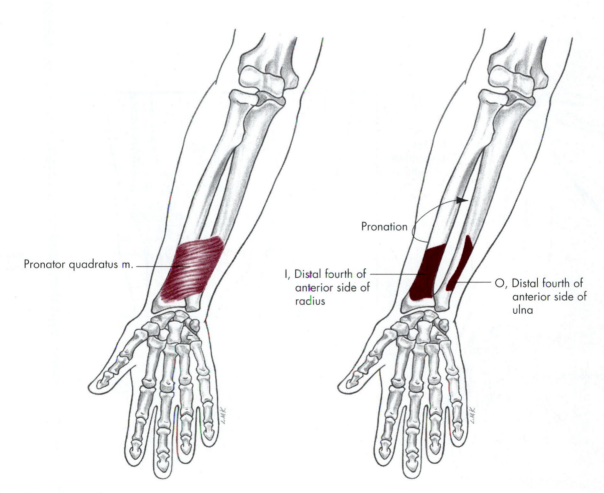

FIG. 4.12 ● Pronator quadratus muscle, anterior view. *O*, Origin; *I*, insertion.

Supinator muscle   FIG. 4.13

(su´pi-na´tor)

Origin

Lateral epicondyle of the humerus and neighboring posterior part of the ulna

Insertion

Lateral surface of the proximal radius just below the head

Action

Supination of the forearm

Palpation

Cannot be palpated

Innervation

Radial nerve (C6)

Application, strengthening, and flexibility

The supinator muscle is called into play when movements of extension and supination are required, such as when turning a screwdriver. The curve in throwing a baseball calls this muscle into play as the elbow is extended just before ball release. It is most isolated in activities that require supination with elbow extension, because the biceps brachii assist with supination most when the elbow is flexed.

The hands should be grasped and the forearms extended, in an attempt to supinate the forearms against the grip of the hands. This localizes, to a degree, the action of the supinator.

Strengthening this muscle begins with holding a hammer in the hand with the hammer head suspended from the ulnar side of the hand while the forearm is supported on a desk or table. The hammer should be hanging toward the floor, with the forearm supinated to the palm-up position.

The supinator is stretched when the forearm is maximally pronated.

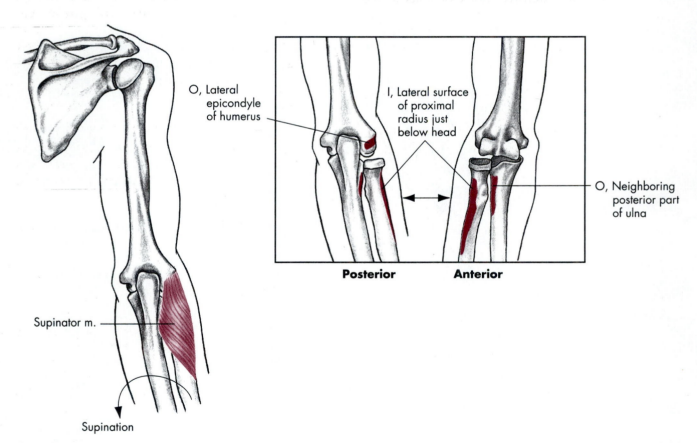

FIG. 4.13 • Supinator muscle, posterior view. *O*, Origin; *I*, insertion.

Web sites

Anatomy & Physiology Tutorials:

www.gwc.maricopa.edu/class/bio201/index.htm

Radiologic Anatomy Browser:

radlinux1.usuf1.usuhs.mil/rad/iong/index.html

This site has numerous radiological views of the musculoskeletal system.

University of Arkansas Medical School Gross Anatomy for Medical Students:

anatomy.uams.edu/htmlpages/anatomyhtml/gross.html

Dissections, anatomy tables, atlas images, links, etc.

Loyola University Medical Center: Structure of the Human Body:

www.meddean.luc.edu/lumen/MedEd/GrossAnatomy/GA.html

An excellent site with many slides, dissections, tutorials, etc., for the study of human anatomy.

Wheeless' Textbook of Orthopaedics:

www.medmedia.com/

This site has an extensive index of links to the fractures, joints, muscles, nerves, trauma, medications, medical topics, and lab tests, as well as links to orthopedic journals and other orthopedic and medical news.

The University of Texas MD Anderson Cancer Center Multimedia and Learning Resources:

rpiwww.mdacc.tmc.edu/mmlearn/anatomy.html

This site has numerous cadaveric cuts of the foot, knee, hand, arm, and elbow; an interactive ankle; and a rotating foot and ankle.

Arthroscopy.Com:

www.arthroscopy.com/sports.htm

Patient information on various musculoskeletal problems of the upper and lower extremity

Premiere Medical Search Engine:

www.medsite.com

This site allows the reader to enter any medical condition and it will search the net to find relevant articles.

Virtual Hospital:

www.vh.org

Numerous slides, patient information, etc.

Electronic Textbook of Hand Surgery:

www.e-hand.com/default.htm

Slides and illustrations of upper extremity musculoskeletal anatomy

Dynamic Human version 2.0 CD-ROM: The Visual Guide to Anatomy & Physiology:

www.mhhe.com/biosci/ap/dynamichuman2/

Web site that accompanies this CD-ROM

Dynamic Human CD activities

1. Review the anatomical landmarks as well as origins for the biceps brachii and triceps brachii. To do so, click on **skeletal, anatomy, gross anatomy, pectoral girdle**, and finally **scapula**.
2. Review anatomical landmarks as well as origins/insertions for the humerus, radius, and ulna as they pertain to the shoulder joint by clicking on **skeletal**; then **anatomy**; then **gross anatomy**, **upper limbs**, and **humerus**; and then **radius/ulna**.
3. Review each of the muscles from this chapter by clicking on **muscular** and then **anatomy, body regions**, and **pectoral girdle** and **upper arm**.
4. Review each of the muscles from this chapter by clicking on **muscular** and then **anatomy, body regions**, and **forearm**.

Worksheet exercises

As an aid to learning, for in-class and out-of-class assignments, or for testing, tear-out worksheets are found at the end of the text (pp. 256 and 257).

Skeletal worksheet (no. 1)

Draw and label on the worksheet the following muscles:
a. Biceps brachii
b. Brachioradialis
c. Brachialis
d. Pronator teres
e. Pronator quadratus
f. Supinator
g. Triceps brachii
h. Anconeus

Human figure worksheet (no. 2)

Label and indicate with arrows the following movements of the elbow and radioulnar joints:
1. Elbow joint
   a. Flexion
   b. Extension
2. Radioulnar joint
   a. Pronation
   b. Supination

Laboratory and review exercises

1. Locate the following parts of the humerus, radius, and ulna on a human skeleton and on a subject.
   a. Skeleton
      1. Medial epicondyle
      2. Lateral epicondyle
      3. Trochlea
      4. Capitulum
      5. Olecranon fossa
      6. Olecranon process
      7. Coronoid process
      8. Coronoid fossa
      9. Tuberosity of the radius
   b. Subject
      1. Medial epicondyle
      2. Lateral epicondyle
      3. Olecranon process
      4. Olecranon fossa
2. How and where can the following muscles be palpated on a human subject?
   a. Biceps brachii
   b. Brachioradialis
   c. Brachialis
   d. Pronator teres
   e. Supinator
   f. Triceps brachii
   g. Anconeus
3. Palpate and list the muscles primarily responsible for the following movements as you demonstrate each:
   a. Flexion
   b. Extension
   c. Pronation
   d. Supination
4. List the planes in which each of the following elbow and radioulnar joint movements occurs. List the respective axis of rotation for each movement in each plane.
   a. Flexion
   b. Extension
   c. Pronation
   d. Supination
5. Discuss the difference between chinning with the palms toward the face and chinning with the palms away from the face. Consider this muscularly and anatomically.
6. Analyze and list the difference in elbow and radioulnar joint muscle activity between turning a door knob clockwise and pushing the door open, versus turning the knob counterclockwise, pulling the door open.
7. Fill in the antagonistic muscle action chart by listing the muscle(s) or parts of muscles that are antagonist in their actions to the muscles in the left column.
8. Fill in the muscle analysis chart by listing the muscles primarily involved in each movement.

Antagonistic muscle action chart • Elbow and radioulnar joints

Agonist	Antagonist
Biceps brachii	
Brachioradialis	
Brachialis	
Pronator teres	
Supinator	
Triceps brachii	
Anconeus	

Muscle analysis chart • Elbow and radioulnar joints

Elbow and radioulnar joints	
Flexion	Extension
Pronation	Supination

References

Andrews JR, Wilk KE: *The athlete's shoulder,* New York, 1994, Churchill Livingstone.

Andrews JR, Zarins B, Wilk KE: *Injuries in baseball,* Philadelphia, 1998, Lippincott-Raven.

Back BR Jr, et al: Triceps rupture: a case report and literature review, *American Journal of Sports Medicine* 15:285, May-June 1987.

Gabbard CP, et al: Effects of grip and forearm position on flex arm hang performance, *Research Quarterly for Exercise and Sport,* July 1983.

Herrick RT, Herrick S: Ruptured triceps in powerlifter presenting as cubital tunnel syndrome—a case report, *American Journal of Sports Medicine* 15:514, September-October 1987.

Hislop HJ, Montgomery J: *Daniels and Worthingham's muscle testing: techniques of manual examination,* ed 6, Philadelphia, 1995, Saunders.

Rasch PJ: *Kinesiology and applied anatomy,* ed 7, Philadelphia, 1989, Lea & Febiger.

Sieg KW, Adams SP: *Illustrated essentials of musculoskeletal anatomy,* ed 2, Gainesville, FL, 1985, Megabooks.

Sisto DJ, et al: An electromyographic analysis of the elbow in pitching, *American Journal of Sports Medicine* 15:260, May-June 1987.

Smith LK, Weiss EL, Lehmkuhl LD: *Brunnstrom's clinical kinesiology,* ed 5, Philadelphia, 1996, Davis.

Springer SI: Racquetball and elbow injuries, *National Racquetball* 16:7, March 1987.

Van De Graaff KM: *Human anatomy,* ed 4, 1995, McGraw-Hill Companies, Inc., New York.

The wrist and hand joints

Objectives

● To identify on a human skeleton selected bony features of the wrist, hand, and fingers

● To label selected bony features on a skeletal chart

● To draw and label the muscles on a skeletal chart

● To palpate the muscles on a human subject while demonstrating their actions

● To list the planes of motion and their respective axes of rotation

● To organize and list the muscles that produce the primary movements of the wrist, hand, and fingers

The joints of the wrist, hand, and fingers are often overlooked in their importance to us in comparison with the larger joints needed for ambulation. Even though the fine motor skills characteristic of this area are not essential in some sports, many sports with skilled activities require precise functioning of the wrist and hand. Several sports, such as archery, bowling, golf, baseball, and tennis, require the combined use of all of these joints.

Because of numerous muscles, bones, and ligaments, combined with relatively small joint size, the functional anatomy of the wrist and hand is complex and overwhelming to some. This complexity may be simplified by relating the functional anatomy to the major actions of the joints: flexion, extension, abduction, and adduction of the wrist and hand.

A large number of muscles are used in these movements. Anatomically and structurally, the human wrist and hand have highly developed, complex mechanisms capable of a variety of movements, which is a result of the arrangement of the 29 bones, more than 25 joints, and more than 30 muscles, of which 18 are intrinsic (both origin and insertion found in the hand) muscles.

For most students who use this text, an extensive knowledge of these intrinsic muscles is not necessary. However, athletic trainers, physical therapists, occupational therapists, chiropractors, anatomists, physicians, and nurses require a more extensive knowledge. The intrinsic muscles are listed, illustrated, and discussed to a limited degree at the end of this chapter. References at the end of this chapter provide additional sources from which to gain further information.

Our discussion is limited to a review of the muscles, joints, and movements involved in gross motor activities. The muscles included are those of the forearm and the extrinsic muscles of the wrist, hand, and fingers. The larger, more important extrinsic muscles of each joint are included, providing a limited knowledge of this area. The prescription of exercises for strengthening these muscles will be somewhat redundant, since there are primarily only four movements accomplished by their combined actions. An exercise that will strengthen most of these muscles is fingertip push-ups.

Bones

The wrist and hand contain 29 bones, including the radius and ulna (Fig. 5.1). Eight carpal bones in two rows of four bones form the wrist. Five metacarpal bones, numbered one to five from the thumb to the little finger, join the wrist bones. There are 14 phalanges (digits), three for each phalange except the thumb, which has only two. They are indicated as proximal, middle, and distal from the metacarpals. Additionally, the thumb has a sesamoid bone within its flexor tendon, and other sesamoids may occur in the fingers.

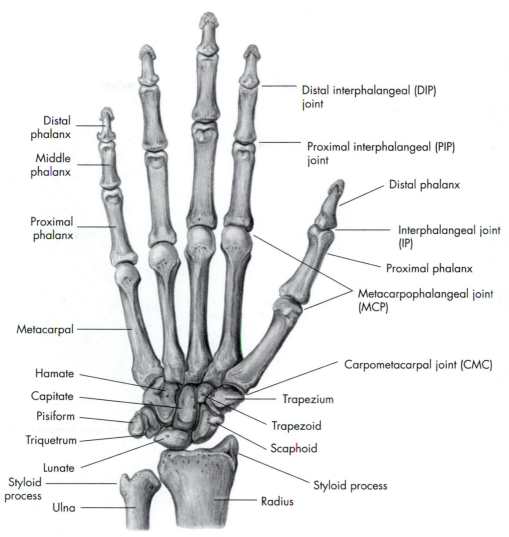

Distal interphalangeal (DIP) joint

Proximal interphalangeal (PIP) joint

Distal phalanx

Interphalangeal joint (IP)

Proximal phalanx

Metacarpophalangeal joint (MCP)

Carpometacarpal joint (CMC)

Trapezium

Trapezoid

Scaphoid

Styloid process

Radius

Distal phalanx

Middle phalanx

Proximal phalanx

Metacarpal

Hamate

Capitate

Pisiform

Triquetrum

Lunate

Styloid process

Ulna

FIG. 5.1 ● Right wrist and hand, palmar surface.

From Anthony CP, Kolthoff NJ: *Textbook of anatomy and physiology*, ed 9, St. Louis, 1975, Mosby.

Joints

The wrist joint is classified as a condyloid-type joint, allowing flexion, extension, abduction, and adduction (Fig. 5.2). Wrist motion occurs primarily between the distal radius and the proximal carpal row, consisting of the scaphoid, lunate, and triquetrum. The joint allows 70 to 90 degrees of flexion and 65 to 85 degrees of extension. The wrist can abduct 15 to 25 degrees and adduct 25 to 40 degrees.

Each finger has three joints. The metacarpophalangeal (MCP) joints are classified as condyloid. In these joints, 0 to 40 degrees of extension and 85 to 100 degrees of flexion are possible. The proximal interphalangeal (PIP) joints, classified as ginglymus, can move from full extension to approximately 90 to 120 degrees of flexion.

The distal interphalangeal (DIP) joints, also classified as ginglymus, can flex 80 to 90 degrees from full extension.

The thumb has only two joints, both of which are classified as ginglymus. The metacarpophalangeal (MCP) joint moves from full extension into 40 to 90 degrees of flexion. The interphalangeal (IP) joint can flex 80 to 90 degrees. The carpometacarpal (CMC) joint of the thumb is a unique saddle-type joint having 50 to 70 degrees of abduction. It can flex approximately 15 to 45 degrees and extend 0 to 20 degrees.

Ligaments, too numerous to mention in this discussion, support and provide static stability to the many joints of the wrist and hand. Some of the finger ligaments are detailed in Fig. 5.3.

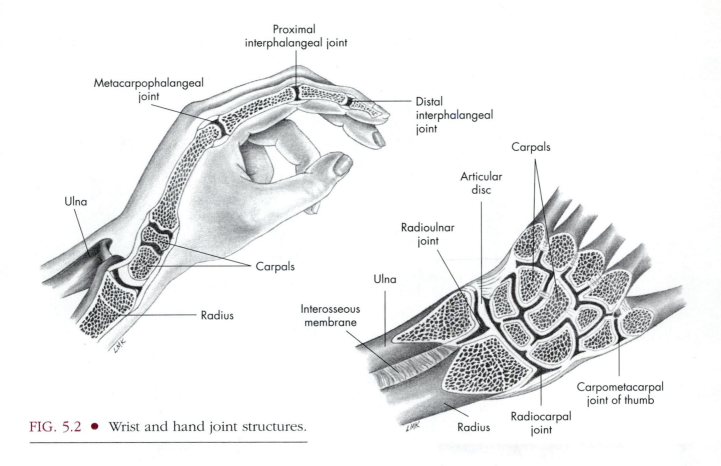

FIG. 5.2 • Wrist and hand joint structures.

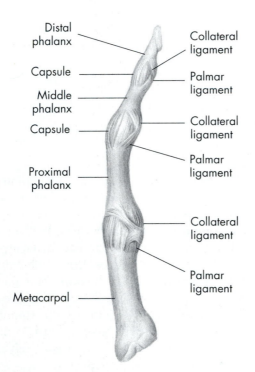

FIG. 5.3 • Metacarpophalangeal and interphalangeal joints, lateral view.

Modified from Van De Graaff KM: *Human Anatomy*, ed 4, Dubuque, IA, 1955, McGraw-Hill Companies, Inc., New York.

Movements

The common actions of the wrist are flexion, extension, abduction, and adduction (Fig. 5.4, *A–D*). The fingers can only flex and extend (Fig. 5.4, *E–F*), except at the metacarpophalangeal joints, where abduction and adduction are controlled by the intrinsic hand muscles. In the hand, the middle phalange is regarded as the reference point by which to differentiate abduction and adduction. Abduction of the index and middle fingers occurs when they move laterally toward the radial side of the forearm. Abduction of the ring and little fingers occurs when they move medially toward the ulnar side of the hand. Movement medially of the index and middle fingers toward the ulnar side of the forearm is adduction. Ring and little finger adduction occurs when these fingers move laterally toward the radial side of the hand. The thumb is abducted when it moves away from the palm and is adducted when it moves toward the palmar aspect of the second metacarpal. These movements, together with pronation and supination of the forearm, make possible the many fine, coordinated movements of the forearm, wrist, and hand.

FIG. 5.4 ● Wrist and hand movements. **A,** Wrist flexion; **B,** wrist extension; **C,** wrist abduction; **D,** wrist adduction; **E,** flexion of the fingers and thumb, adduction of metacarpophalangeal joints, opposition of thumb; **F,** extension of the fingers and thumb, abduction of metacarpophalangeal joints.

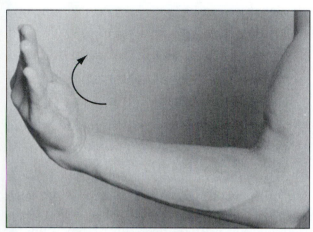

A

Wrist flexion

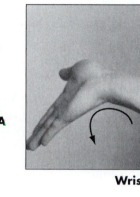

B

Wrist extension

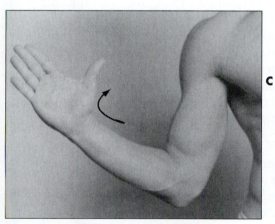

C

Wrist abduction (radial flexion)

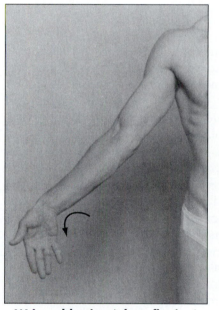

D

Wrist adduction (ulnar flexion)

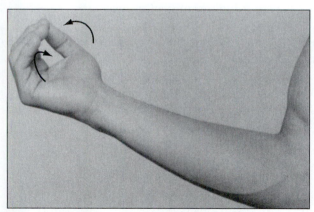

E

Flexion of fingers and thumb, adduction of metacarpophalangeal joints, opposition of thumb

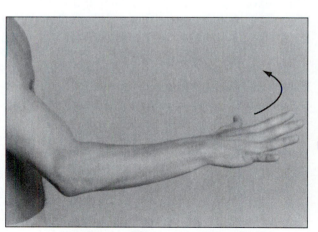

F

Extension of fingers and thumb, abduction of metacarpophalangeal joints

Flexion: movement of the palm of the hand and/or the phalanges toward the anterior or volar aspect of the forearm

Extension: movement of the back of the hand and/or the phalanges toward the posterior or dorsal aspect of the forearm; sometimes referred to as hyperextension

Abduction: (radial flexion) movement of the thumb side of the hand toward the lateral aspect or radial side of the forearm; also, movement of the fingers away from the middle finger

Adduction: (ulnar flexion) movement of the little finger side of the hand toward the medial aspect or ulnar side of the forearm; also, movement of the fingers back together toward the middle finger

Opposition: movement of the thumb across the palmar aspect to oppose any or all of the phalanges

Muscles

The extrinsic muscles of the wrist and hand can be grouped according to function and location. There are six muscles that move the wrist but do not cross the hand to move the fingers and thumb. The three wrist flexors in this group are the flexor carpi radialis, flexor carpi ulnaris, and palmaris longus—all of which have their origin on the medial epicondyle of the humerus. The extensors of the wrist have their origin on the lateral epicondyle and include the extensor carpi radialis longus, extensor carpi radialis brevis, and extensor carpi ulnaris.

Another nine muscles function primarily to move the phalanges but are also involved in wrist joint actions because they originate on the forearm and cross the wrist. These muscles generally are weaker in their actions on the wrist. The flexor digitorum superficialis and the flexor digitorum profundus are finger flexors; however, they also assist in wrist flexion along with the flexor pollicis longus, which is a thumb flexor. The extensor digitorum, the extensor indicis, and the extensor digiti minimi are finger extensors but also assist in wrist extension, along with the extensor pollicis longus and the extensor pollicis brevis, which extend the thumb. The abductor pollicis longus abducts the thumb and assists in wrist abduction.

All of the wrist flexors generally have their origins on the anteromedial aspect of the proximal forearm and medial epicondyle of the humerus, whereas their insertions are on the anterior aspect of the wrist and hand. The wrist extensors generally have their origins on the posterolateral aspect of the proximal forearm and lateral humeral epicondyle, whereas their insertions are located on the posterior aspect of the wrist and hand.

The wrist abductors are the flexor carpi radialis, extensor carpi radialis longus, extensor carpi radialis brevis, abductor pollicis longus, extensor pollicis longus, and extensor pollicis brevis. These muscles generally cross the wrist joint anterolaterally and posterolaterally to insert on the radial side of the hand. The flexor carpi ulnaris and extensor carpi ulnaris adduct the wrist and cross the wrist joint anteromedially and posteromedially to insert on the ulnar side of the hand.

The intrinsic muscles of the hand have their origins and insertions on the bones of the hand. Grouping of the intrinsic muscles into three groups according to location is helpful in understanding and learning these muscles. On the radial side there are four muscles of the thumb—the opponens pollicis, the abductor pollicis brevis, the flexor pollicis brevis, and the adductor pollicis. On the ulnar side, there are three muscles of the little finger—the opponens digiti minimi, the abductor digiti minimi, and the flexor digiti minimi brevis. In the remainder of the hand, there are 11 muscles, which can be further grouped as the four lumbricals, the three palmar interossei, and the four dorsal interossei.

Wrist and hand muscles—location

Anteromedially at the elbow and forearm and
    anterior at the hand
        Primarily wrist and hand flexion
            Flexor carpi radialis
            Flexor carpi ulnaris
            Palmaris longus
            Flexor digitorum superficialis
            Flexor digitorum profundus
            Flexor pollicis longus
Posterolaterally at the elbow and forearm and
    posterior at the hand
        Primarily wrist and hand extension
            Extensor carpi radialis longus
            Extensor carpi radialis brevis
            Extensor carpi ulnaris
            Extensor digitorum
            Extensor indicis
            Extensor digiti minimi
            Extensor pollicis longus
            Extensor pollicis brevis
            Abductor pollicis longus

Flexor carpi radialis muscle   FIG. 5.5

(fleks´or kar´pi ra´di-a´lis)

Origin

Medial epicondyle of the humerus

Insertion

Base of the second and third metacarpals, anterior
  (palmar surface)

Action

Flexion of the wrist
Abduction of wrist
Weak flexion of the elbow

Palpation

Anterior surface of the wrist, slightly lateral, in line
  with the second and third metacarpals

Innervation

Median nerve (C6, 7)

Application, strengthening, and flexibility

The flexor carpi radialis, flexor carpi ulnaris, and
palmaris longus are the most powerful of the wrist
flexors. They are brought into play during any activity that requires wrist curling or stabilization of
the wrist against resistance, particularly if the forearm is supinated.

The flexor carpi radialis may be developed by
performing wrist curls against a handheld resistance. This may be accomplished when the
supinated forearm is supported by a table, with
the hand and wrist hanging over the edge to allow
full range of motion. The extended wrist is then
flexed or curled up to strengthen this muscle.

To stretch the flexor carpi radialis, the elbow
must be fully extended with the forearm
supinated while a partner passively extends and
adducts the wrist.

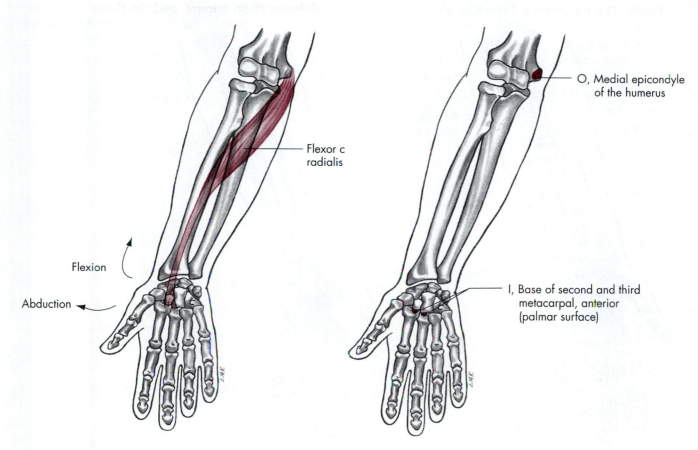

Flexor c
radialis

Flexion

Abduction

O, Medial epicondyle
of the humerus

I, Base of second and third
metacarpal, anterior
(palmar surface)

FIG. 5.5 ● Flexor carpi radialis muscle, anterior view. O, Origin; I, insertion.

Palmaris longus muscle   FIG. 5.6

(pal-ma´ris long´gus)

Origin
Medial epicondyle of the humerus

Insertion
Palmar aponeurosis of the second, third, fourth, and fifth metacarpals

Action
Flexion of the wrist
Weak flexion of the elbow

Palpation
Anterior medial aspect of forearm and central aspect of the anterior forearm just proximal to the wrist

Innervation
Median nerve (C6, 7)

Application, strengthening, and flexibility

Unlike the flexor carpi radialis and flexor carpi ulnaris, which are not only wrist flexors but also abductors and adductors, respectively, the palmaris longus is involved only in wrist flexion because of its central location on the anterior forearm and wrist. It may also be strengthened with any type of wrist-curling activity, such as the ones described for the flexor carpi radialis muscle.

• Maximal elbow and wrist extension stretches the palmaris longus.

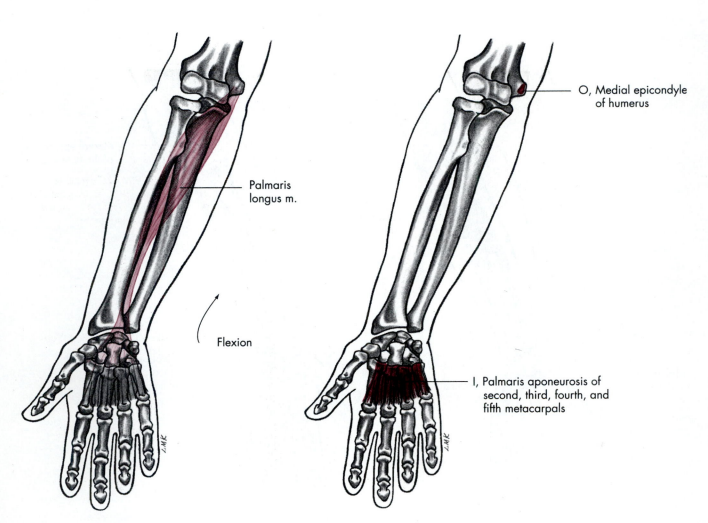

Palmaris longus m.

Flexion

O, Medial epicondyle of humerus

I, Palmaris aponeurosis of second, third, fourth, and fifth metacarpals

FIG. 5.6 • Palmaris longus muscle, anterior view. *O*, Origin; *I*, insertion.

Flexor carpi ulnaris muscle FIG. 5.7
(fleks´or kar´pi ul-na´ris)

Origin

Medial epicondyle of the humerus
Posterior aspect of the proximal ulna

Insertion

Pisiform, hamate, and base of the fifth metacarpal
  (palmar surface)

Action

Flexion of the wrist
Adduction of the wrist, together with the extensor
  carpi ulnaris muscle
Weak flexion of the elbow

Palpation

Anteromedial surface of the forearm, a few inches
  below the medial epicondyle of the humerus to
  just proximal to the wrist

Innervation

Ulnar nerve (C8, T1)

Application, strengthening, and flexibility

The flexor carpi ulnaris is very important in wrist
flexion or curling activities. In addition, it is one of
only two muscles involved in wrist adduction or
ulnar flexion. It may be strengthened with any
type of wrist-curling activity against resistance
such as those described for the flexor carpi radialis muscle.

To stretch the flexor carpi ulnaris, the elbow
must be fully extended with the forearm
supinated while a partner passively extends and
abducts the wrist.

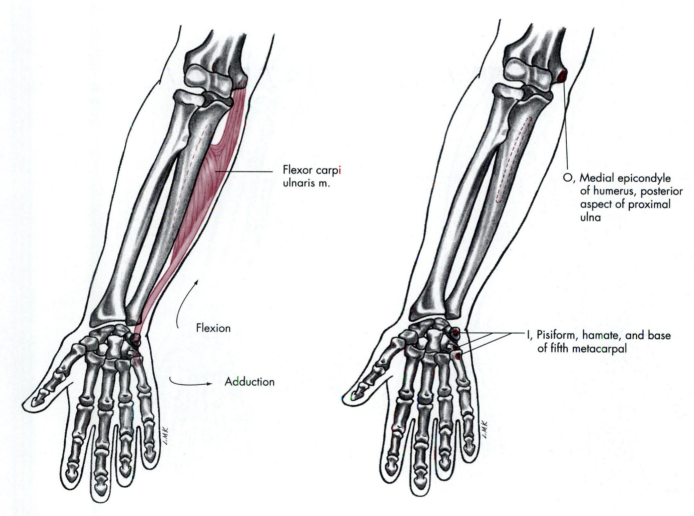

Flexor carpi
ulnaris m.

Flexion

Adduction

O, Medial epicondyle
of humerus, posterior
aspect of proximal
ulna

I, Pisiform, hamate, and base
of fifth metacarpal

FIG. 5.7 ● Flexor carpi ulnaris muscle, anterior view. *O*, Origin; *I*, insertion.

Extensor carpi ulnaris muscle   FIG. 5.8

(eks-ten´sor kar´pi ul-na´ris)

Origin

Lateral epicondyle of the humerus
Middle two-fourths of the posterior border of
  the ulna

Insertion

Base of the fifth metacarpal (dorsal surface)

Action

Extension of the wrist
Adduction of the wrist together with the flexor
  carpi ulnaris muscle
Weak extension of the elbow

Palpation

Anterior ulnar side of the forearm near the fifth
  metacarpal

Innervation

Radial nerve (C6–8)

Application, strengthening, and flexibility

Besides being a powerful wrist extensor, the extensor carpi ulnaris muscle is the only muscle other than the flexor carpi ulnaris involved in wrist adduction or ulnar flexion. Wrist extension exercises, such as those described for the extensor carpi radialis longus, are appropriate for development of the muscle.

Stretching the extensor carpi ulnaris requires the elbow to be extended with the forearm pronated while the wrist is passively flexed and slightly abducted.

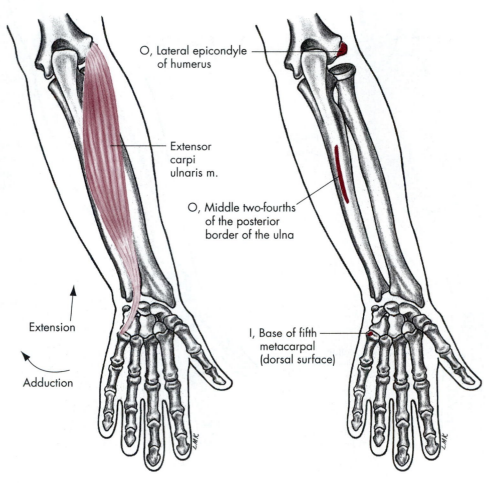

O, Lateral epicondyle
of humerus

Extensor
carpi
ulnaris m.

O, Middle two-fourths
of the posterior
border of the ulna

Extension

Adduction

I, Base of fifth
metacarpal
(dorsal surface)

FIG. 5.8 ● Extensor carpi ulnaris muscle, posterior view. *O*, Origin; *I*, insertion.

Extensor carpi radialis brevis muscle

FIG. 5.9

(eks-ten´sor kar´pi ra´di-a´lis bre´vis)

Origin
Lateral epicondyle of the humerus

Insertion
Base of the third metacarpal (dorsal surface)

Action
Extension of the wrist
Abduction of the wrist
Weak extension of the elbow

Palpation
Dorsal side of the forearm, which is difficult to palpate

Innervation
Radial nerve (C6, 7)

Application, strengthening, and flexibility

The extensor carpi radialis brevis, like the extensor carpi radialis longus, is important in any sports activity that requires powerful wrist extension, such as golf or tennis. In addition, both of these muscles are involved in abduction of the wrist. The extensor carpi radialis brevis may be developed with the same wrist extension exercises as described for the extensor carpi radialis longus muscle.

Stretching the extensor carpi radialis brevis and longus requires the elbow to be extended with the forearm pronated while the wrist is passively flexed and slightly adducted.

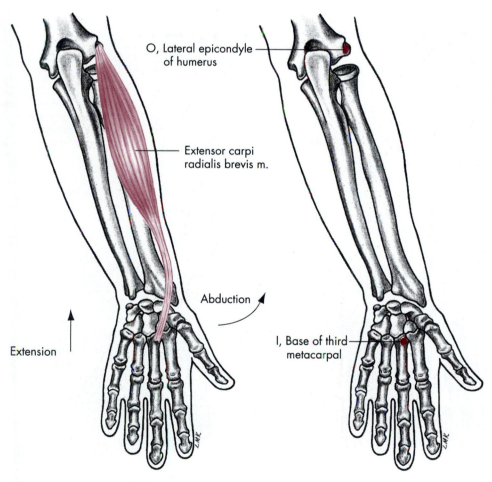

O, Lateral epicondyle of humerus

Extensor carpi radialis brevis m.

Extension

Abduction

I, Base of third metacarpal

FIG. 5.9 ● Extensor carpi radialis brevis muscle, posterior view. O, Origin; I, insertion.

Extensor carpi radialis longus muscle

FIG. 5.10

(eks-ten´sor kar´pi ra´di-a´lis long´gus)

Origin

Lower third of lateral supracondylar ridge of
humerus and lateral epicondyle of the humerus

Insertion

Base of the second metacarpal (dorsal surface)

Action

Extension of the wrist
Abduction of the wrist
Weak extension of the elbow

Palpation

Posterior aspect of proximal forearm and anterolateral surface just proximal to the wrist, in line with
the second metacarpal

Innervation

Radial nerve (C6, 7)

Application, strengthening, and flexibility

The extensor carpi radialis longus, extensor carpi
radialis brevis, and extensor carpi ulnaris are the
most powerful of the wrist extensors. Any activity
requiring wrist extension or stabilization of the
wrist against resistance, particularly if the forearm
is pronated, depends greatly on the strength of
these muscles. They are often brought into play
with the backhand in racket sports.

The extensor carpi radialis longus may be developed by performing wrist extension against a
handheld resistance. This may be accomplished
with the pronated forearm being supported by a
table, with the hand hanging over the edge to
allow full range of motion. The wrist is then
moved from the fully flexed position to the fully
extended position against the resistance.

The extensor carpi radialis longus is stretched
in the same manner as the extensor carpi radialis
brevis.

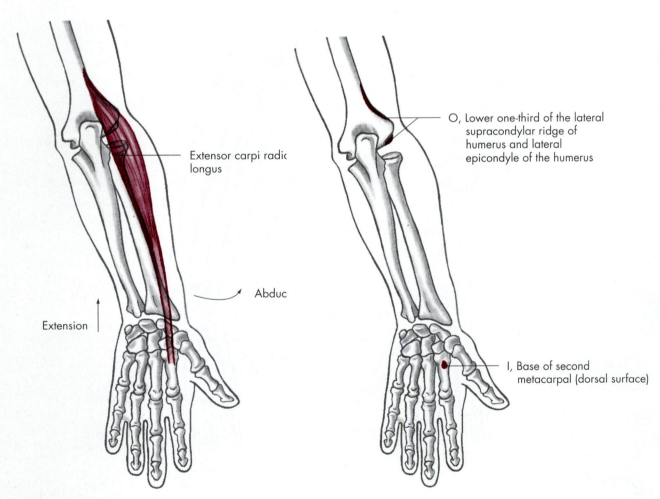

Extensor carpi radic longus

Abduc

Extension

O, Lower one-third of the lateral
supracondylar ridge of
humerus and lateral
epicondyle of the humerus

I, Base of second
metacarpal (dorsal surface)

FIG. 5.10 • Extensor carpi radialis longus muscle, posterior view. *O,* Origin; *I,* insertion.

Flexor digitorum superficialis muscle

FIG. 5.11

(fleks´or dij-i-to´rum su´per-fish-e-al´is)

Origin

Medial epicondyle of the humerus
Ulnar head: medial coronoid process
Radial head: upper two-thirds of anterior border of
   radius

Insertion

Each tendon splits and attaches to the sides of the
   middle phalanx of four fingers (palmar surface)

Action

Flexion of the fingers at the metacarpophalangeal
   and proximal interphalangeal joints
Flexion of the wrist
Weak flexion of the elbow

Palpation

Anterior wrist surface on the ulnar side next to the
   flexor carpi ulnaris muscle

Innervation

Median nerve (C7, 8, and T1)

Application, strengthening, and flexibility

The flexor digitorum superficialis muscle, also
known as the flexor digitorum sublimis, divides
into four tendons on the palmar aspect of the wrist
and hand to insert on each of the four fingers. The
flexor digitorum superficialis and the flexor digitorum profundus are the only muscles involved in
flexion of all four fingers. Both of these muscles
are vital in any type of gripping activity.

Squeezing a sponge rubber ball in the palm of
the hand, along with other gripping and squeezing
activities, can be used to develop these muscles.

The flexor digitorum superficialis is stretched
by passively extending the elbow, wrist, metacarpophalangeal, and proximal interphalangeal joints
while maintaining the forearm in full supination.

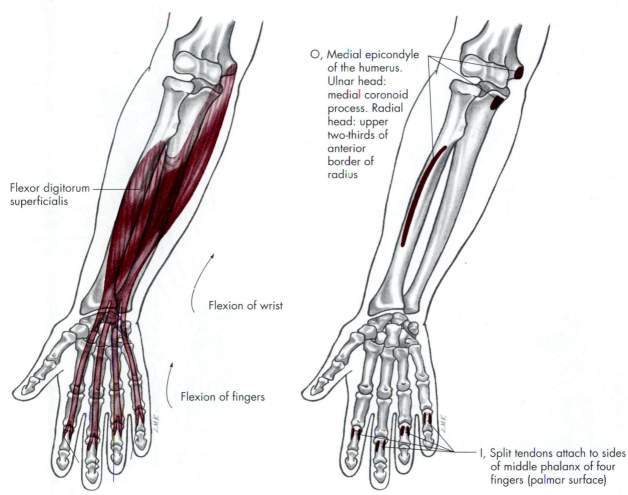

O, Medial epicondyle
of the humerus.
Ulnar head:
medial coronoid
process. Radial
head: upper
two-thirds of
anterior
border of
radius

Flexor digitorum
superficialis

Flexion of wrist

Flexion of fingers

I, Split tendons attach to sides
of middle phalanx of four
fingers (palmar surface)

FIG. 5.11 ● Flexor digitorum superficialis muscle, anterior view. O, Origin; I, insertion.

Flexor digitorum profundus muscle

FIG. 5.12

(fleks´or dij-i-to´rum pro-fun´dus)

Origin

Proximal three-fourths of the anterior and medial ulna

Insertion

Base of the distal phalanxes of the four fingers

Action

Flexion of the four fingers at the metacarpophalangeal, proximal interphalangeal, and distal interphalangeal joints
Flexion of the wrist

Palpation

Anterior surfaces of the middle phalanges of the four fingers

Innervation

Median nerve (C8, T1) to the second and third fingers
Ulnar nerve (C8, T1) to the fourth and fifth fingers

Application, strengthening, and flexibility

Both the flexor digitorum profundus muscle and the flexor digitorum superficialis muscle assist in wrist flexion because of their palmar relationship to the wrist. The flexor digitorum profundus is used in any type of gripping, squeezing, or hand-clenching activity, such as gripping a racket or climbing a rope.

The flexor digitorum profundus muscle may be developed through these activities, in addition to the strengthening exercises described for the flexor digitorum superficialis muscle.

The flexor digitorum profundus is stretched similarly to the flexor digitorum superficialis, except that the distal interphalangeal joints must be passively extended in addition to the elbow, wrist, metacarpophalangeal, and proximal interphalangeal joints while maintaining the forearm in full supination.

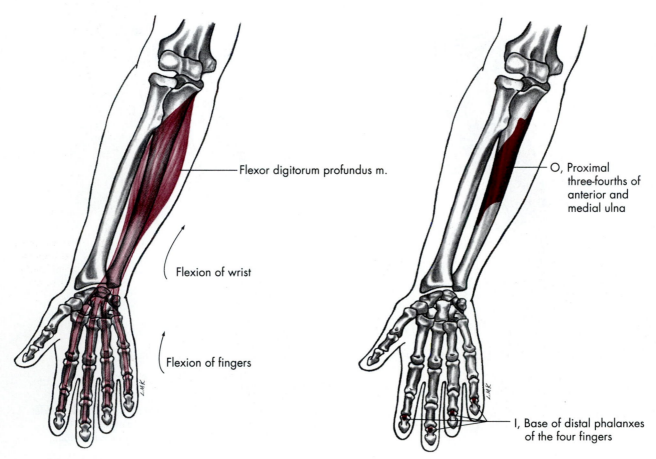

Flexor digitorum profundus m.

Flexion of wrist

Flexion of fingers

O, Proximal three-fourths of anterior and medial ulna

I, Base of distal phalanxes of the four fingers

FIG. 5.12 • Flexor digitorum profundus muscle, anterior view. *O*, Origin; *I*, insertion.

Flexor pollicis longus muscle   FIG. 5.13

(fleks´or pol´i-sis long´gus)

Origin

Middle anterior surface of the radius and the anterior medial border of the ulna just distal to the coronoid process

Insertion

Base of the distal phalanx of the thumb (palmar surface)

Action

Flexion of the thumb carpometacarpal, metacarpophalangeal, and interphalangeal joints
Flexion of the wrist

Palpation

Anterior surface of the thumb

Innervation

Median nerve, palmar interosseous branch (C8, T1)

Application, strengthening, and flexibility

The primary function of the flexor pollicis longus muscle is flexion of the thumb, which is vital in gripping and grasping activities of the hand. Because of its palmar relationship to the wrist, it provides some assistance in wrist flexion.

It may be strengthened by pressing a sponge rubber ball into the hand with the thumb and by many other gripping or squeezing activities.

The flexor pollicis longus is stretched by passively extending the entire thumb while simultaneously maintaining maximal wrist extension.

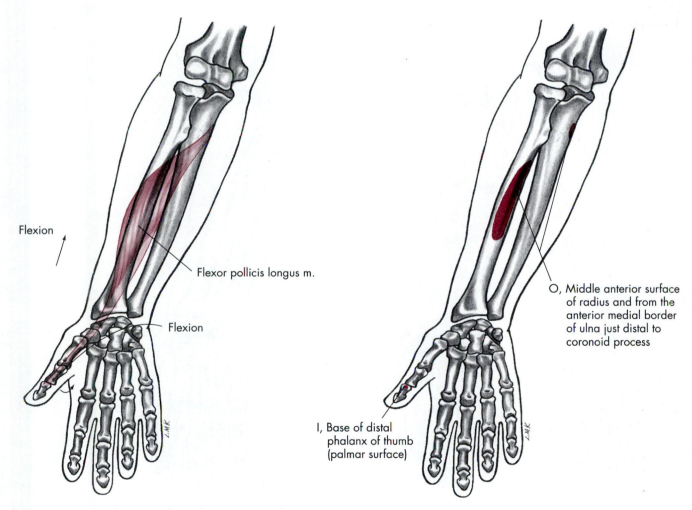

Flexion

Flexor pollicis longus m.

Flexion

O, Middle anterior surface of radius and from the anterior medial border of ulna just distal to coronoid process

I, Base of distal phalanx of thumb (palmar surface)

FIG. 5.13 ● Flexor pollicis longus muscle, anterior view. *O*, Origin; *I*, insertion.

Extensor digitorum muscle   FIG. 5.14

(eks-ten´sor dij-i-to´rum)

Origin

Lateral epicondyle of the humerus

Insertion

Four tendons to bases of middle and distal
  phalanges of four fingers (dorsal surface)

Action

Extension of the second, third, fourth, and fifth
  phalanges at the metacarpophalangeal joints
Extension of the wrist
Weak extension of the elbow

Palpation

Middorsal surface of the forearm and dorsal aspect
  of the hand

Innervation

Radial nerve (C6–8)

Application, strengthening, and flexibility

The extensor digitorum, also known as the extensor digitorum communis, is the only muscle involved in extension of all four fingers. This muscle divides into four tendons on the dorsum of the
wrist to insert on each of the fingers. It also assists
with wrist extension movements. It may be developed by applying manual resistance to the dorsal
aspect of the flexed fingers and then extending
the fingers fully. When performed with the wrist
in flexion, this exercise increases the workload on
the extensor digitorum.

To stretch the extensor digitorum, the fingers
must be maximally flexed at the metacarpophalangeal, proximal interphalangeal, and distal interphalangeal joints while the wrist is fully
flexed.

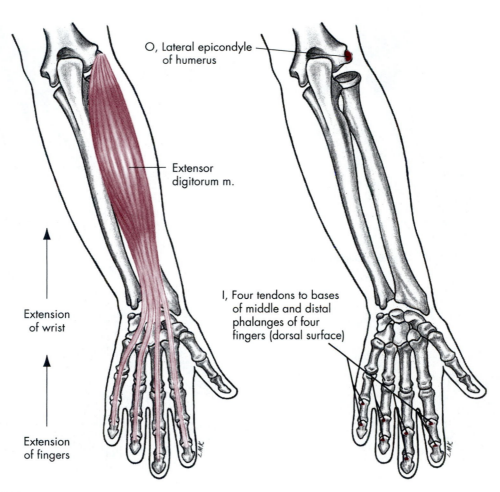

FIG. 5.14 ● Extensor digitorum muscle, posterior view. *O*, Origin; *I*, insertion.

Extensor indicis muscle   FIG. 5.15

(eks-ten´sor in´di-sis)

Origin

Middle to distal one-third of the posterior ulna

Insertion

Base of the middle and distal phalanxes of second phalange (dorsal surface)

Action

Extension of the index finger at the metacarpophalangeal joint

Weak wrist extension

Palpation

Posterior aspect of the distal forearm and dorsal surface of the hand just medial to the extensor digitorum tendon of the index finger

Innervation

Radial nerve (C6–8)

Application, strengthening, and flexibility

The extensor indicis muscle is the pointing muscle. That is, it is responsible for extending the index finger, particularly when the other fingers are flexed. It also provides weak assistance to wrist extension and may be developed through exercises similar to those described for the extensor digitorum.

The extensor indicis is stretched by passively taking the index finger into maximal flexion at its metacarpophalangeal, proximal interphalangeal, and distal interphalangeal joints while fully flexing the wrist.

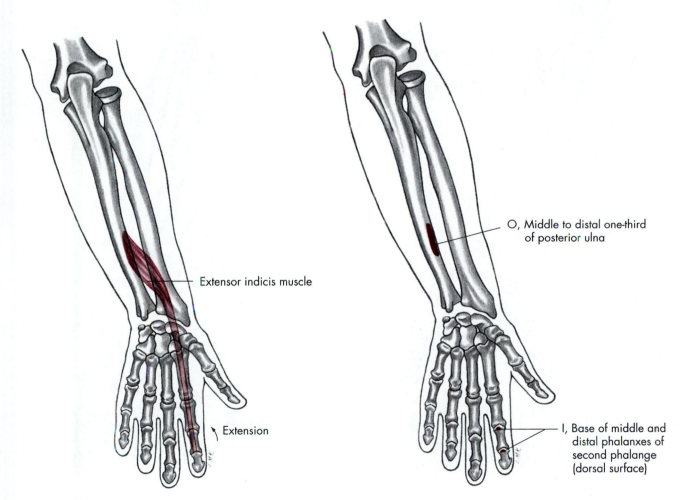

Extensor indicis muscle

Extension

O, Middle to distal one-third of posterior ulna

I, Base of middle and distal phalanxes of second phalange (dorsal surface)

FIG. 5.15 ● Extensor indicis muscle, posterior view. *O,* Origin; *I,* insertion.

Extensor digiti minimi muscle   FIG. 5.16

(eks-ten´sor dij´i-ti min´im-i)

Origin

Lateral epicondyle of the humerus

Insertion

Base of the middle and distal phalanxes of the fifth
  phalange (dorsal surface)

Action

Extension of the little finger at the metacarpophalangeal joint
Weak wrist extension

Palpation

Cannot be palpated

Innervation

Radial nerve (C6–8)

Application, strengthening, and flexibility

The primary function of the extensor digiti minimi
muscle is to assist the extensor digitorum in extending the little finger. Because of its dorsal
relationship to the wrist, it also provides weak assistance in wrist extension. It is strengthened with
the same exercises described for the extensor
digitorum.

The extensor digiti minimi is stretched by passively taking the little finger into maximal flexion
at its metacarpophalangeal, proximal interphalangeal, and distal interphalangeal joints while
fully flexing the wrist.

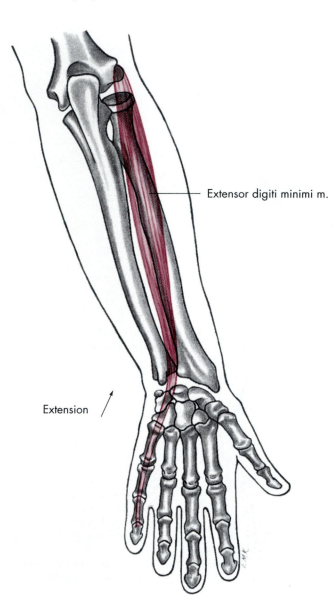

Extensor digiti minimi m.

Extension

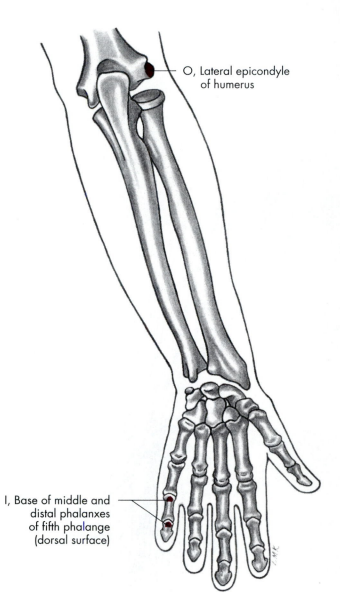

O, Lateral epicondyle
of humerus

I, Base of middle and
distal phalanxes
of fifth phalange
(dorsal surface)

FIG. 5.16 • Extensor digiti minimi muscle, posterior view. *O*, Origin; *I*, insertion.

Extensor pollicis longus muscle

FIG. 5.17

(eks-ten´sor pol´i-sis long´gus)

Origin

Posterior lateral surface of the lower middle ulna

Insertion

Base of the distal phalanx of the thumb (dorsal surface)

Action

Extension of the wrist
Extension of the thumb

Palpation

Most prominent on the dorsal side of the thumb

Innervation

Radial nerve (C6–8)

Application, strengthening, and flexibility

The primary function of the extensor pollicis longus muscle is extension of the thumb, although it does provide weak assistance in wrist extension.

It may be strengthened by extending the flexed thumb against manual resistance. It is stretched by passively taking the entire thumb into maximal flexion at its carpometacarpal, metacarpophalangeal, and interphalangeal joints while fully flexing the wrist.

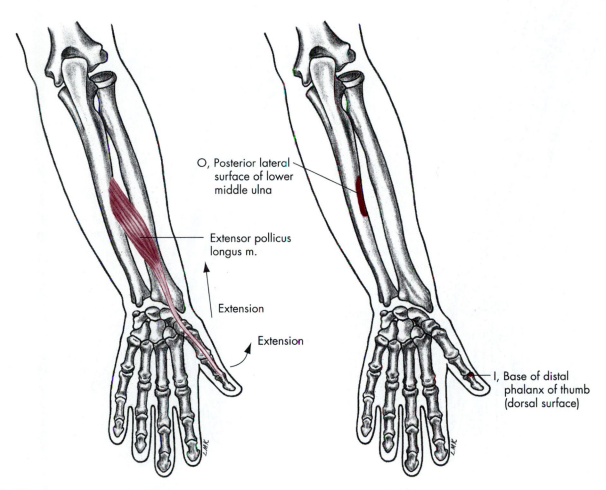

O, Posterior lateral surface of lower middle ulna

Extensor pollicus longus m.

Extension

Extension

I, Base of distal phalanx of thumb (dorsal surface)

FIG. 5.17 ● Extensor pollicis longus muscle, posterior view. *O*, Origin; *I*, insertion.

Extensor pollicis brevis muscle

FIG. 5.18

(eks-ten′sor pol′i-sis bre′vis)

Origin

Posterior surface of the lower middle radius

Insertion

Base of the proximal phalanx of the thumb (dorsal surface)

Action

Extension of the thumb at the metacarpophalangeal joint

Weak wrist extension

Palpation

Just lateral to the extensor pollicis longus tendon on the dorsal side of the hand

Most prominent on the dorsal side of the hand and wrist

Innervation

Radial nerve (C6, 7)

Application, strengthening, and flexibility

The extensor pollicis brevis assists the extensor pollicis longus in extending the thumb. Because of its dorsal relationship to the wrist, it, too, provides weak assistance in wrist extension.

It may be strengthened through the same exercises described for the extensor pollicis longus muscle. It is stretched by passively taking the first carpometacarpal joint and the metacarpophalangeal joint of the thumb into maximal flexion while fully flexing the wrist.

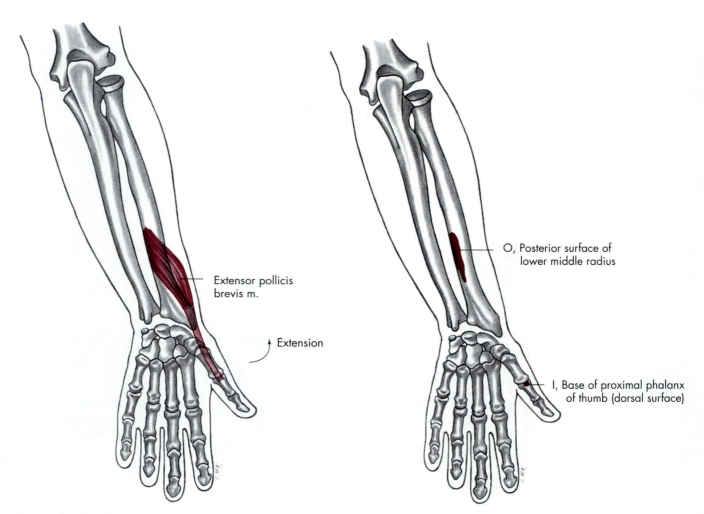

Extensor pollicis brevis m.

Extension

O, Posterior surface of lower middle radius

I, Base of proximal phalanx of thumb (dorsal surface)

FIG. 5.18 ● Extensor pollicis brevis muscle, posterior view. *O*, Origin; *I*, insertion.

Abductor pollicis longus muscle

FIG. 5.19

(ab-duk´tor pol´i-sis lon´gus)

Origin

Posterior aspect of the radius and midshaft of the ulna

Insertion

Base of the first metacarpal (dorsal surface)

Action

Abduction of the thumb at the carpometacarpal joint

Abduction of the wrist

Palpation

Lateral aspect of the wrist joint just proximal to the first metacarpal

Innervation

Radial nerve (C6, 7)

Application, strengthening, and flexibility

The primary function of the abductor pollicis longus muscle is abduction of the thumb, although it does provide some assistance in abduction of the wrist. It may be developed by abducting the thumb from the adducted position against a manually applied resistance. Stretching of the abductor pollicis longus is accomplished by fully flexing and adducting the entire thumb across the palm with the wrist fully adducted.

The abductor pollicis brevis, along with the tendons of the extensor pollicis longus and brevis, forms the anatomical snuffbox, which is the small depression that develops between these two tendons when they contract. The name "anatomical snuffbox" originates from tobacco users' placing their snuff in this depression.

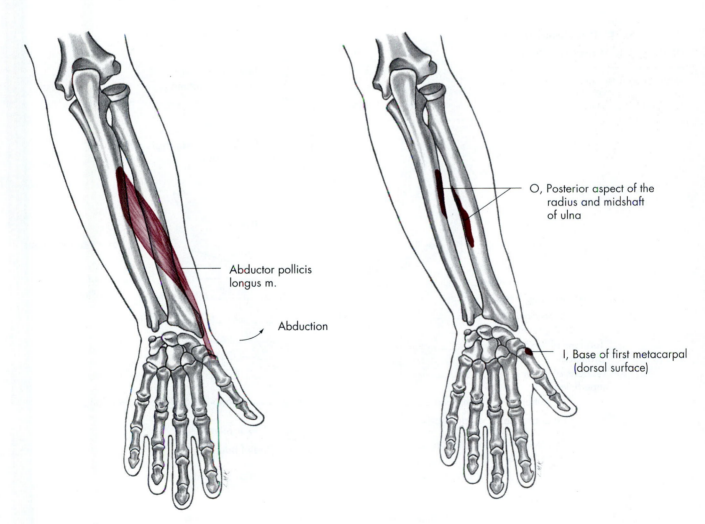

Abductor pollicis longus m.

Abduction

O, Posterior aspect of the radius and midshaft of ulna

I, Base of first metacarpal (dorsal surface)

FIG. 5.19 ● Abductor pollicis longus muscle, posterior view. *O,* Origin; *I,* insertion.

Intrinsic muscles of the hand

The intrinsic hand muscles may be grouped according to location as well as according to the parts of the hand they act (Fig. 5.20). The abductor pollicis brevis, opponens pollicis, flexor pollicis brevis, and adductor pollicis make up the thenar eminence—the muscular pad on the palmar surface of the first metacarpal. The hypothenar eminence is the muscular pad that forms the ulnar border on the palmar surface of the hand and is made up of the abductor digiti minimi, flexor digiti minimi brevis, and opponens digiti minimi. The intermediate muscles of the hand consist of three palmar interossei, four dorsal interossei, and four lumbrical muscles.

Four intrinsic muscles act on the carpometacarpal joint of the thumb. The opponens pollicis is the muscle that causes opposition in the thumb metacarpal. The abductor pollicis brevis abducts the thumb metacarpal and is assisted in this action by the flexor pollicis brevis, which also flexes the thumb metacarpal. The metacarpal of the thumb is adducted by the adductor pollicis. Both the flexor pollicis brevis and the adductor pollicis flex the proximal phalanx of the thumb.

The three palmar interossei are adductors of the second, fourth, and fifth phalanges. The four dorsal interossei both flex and abduct the index, middle, and ring proximal phalanxes, in addition to assisting with extension of the middle and distal phalanxes of these fingers. The third dorsal interossei also adducts the middle finger. The four lumbricals flex the index, middle, ring, and little proximal phalanxes and extend the middle and distal phalanxes of these fingers.

Three muscles act on the little finger. The opponens digiti minimi causes opposition of the little finger metacarpal. The abductor digiti minimi abducts the little finger metacarpal, and the flexor digiti minimi brevis flexes this metacarpal.

Refer to Table 5.1 for further details regarding the intrinsic muscles of the hand.

FIG. 5.20 ● Interossei muscles of the hand, anterior view.

Modified from Van De Graaff KM: *Human anatomy*, ed 4, 1995, McGraw-Hill Companies, Inc., New York.

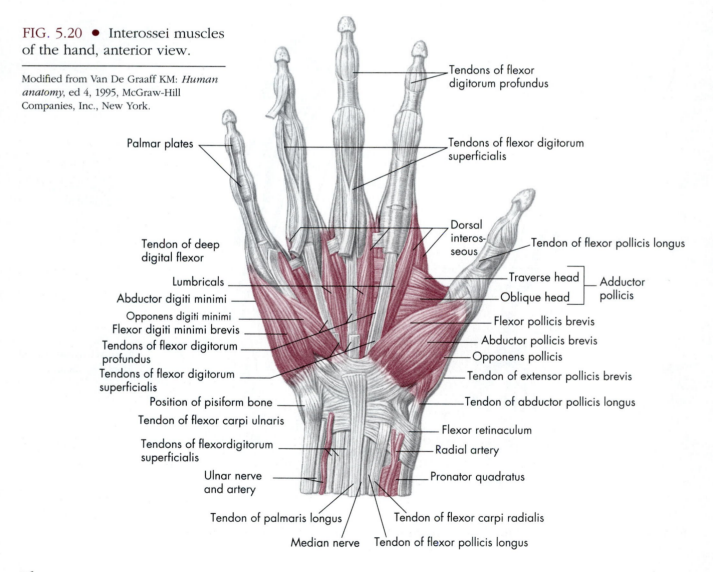

TABLE 5.1 • Intrinsic muscles of the hand

Muscle	Origin	Insertion	Action	Palpation	Innervation
Opponens pollicis	Anterior surface of transverse carpal ligament, trapezium	Lateral border of 1st metacarpal	CMC opposition of thumb	Lateral shaft of 1st metacarpal	Median nerve (C6, 7)
Abductor pollicis brevis	Anterior surface of transverse carpal ligament, trapezium, scaphoid	Base of 1st proximal phalanx	CMC abduction of thumb	Radial aspect of palmar surface of 1st metacarpal	Median nerve (C6, 7)
Flexor pollicis brevis	Superficial head: trapezium and transverse carpal ligament Deep head: ulnar aspect of 1st metacarpal	Base of proximal phalanx of 1st metacarpal	CMC flexion and abduction; MCP flexion of thumb	Ulnar side of thenar eminence just proximal to MCP joint	Superficial head: median nerve (C6, 7) Deep head: ulnar nerve (C8, T1)
Adductor pollicis	Transverse head: anterior shaft of 3rd metacarpal Oblique head: base of 2nd and 3rd metacarpals, capitate, trapezoid	Ulnar aspect of base of proximal phalanx of 1st metacarpal	CMC adduction; MCP flexion of thumb	Palmar surface between 1st and 2nd metacarpal	Ulnar nerve (C8, T1)
Palmar interossei	Shaft of 2nd, 4th, and 5th metacarpals and extensor expansions	Bases of 2nd, 4th, and 5th proximal phalanxes and extensor expansions	MCP adduction of 2nd, 4th, and 5th phalanges	Cannot be palpated	Ulnar nerve (C8, T1)
Dorsal interossei	Two heads on shafts on adjacent metacarpals	Bases of 2nd, 3rd, and 4th proximal phalanxes and extensor expansions	MCP flexion and abduction; PIP/DIP extension of 2nd, 3rd, and 4th phalanges; MCP adduction of 3rd phalange	Dorsal surface between 1st and 2nd metacarpals, between shafts of 2nd through 5th metacarpals	Ulnar nerve, palmar branch (C8, T1)
Lumbricales	Flexor digitorum profundus tendon in center of palm	Extensor expansions on radial side of 2nd, 3rd, 4th, and 5th proximal phalanxes	MCP flexion and PIP/DIP extension of 2nd, 3rd, 4th, and 5th phalanges	Cannot be palpated	1st and 2nd: median nerve (C6, 7) 3rd and 4th: ulnar nerve (C8, T1)
Opponens digiti minimi	Hook of hamate and adjacent transverse carpal ligament	Medial border of 5th metacarpal	MCP opposition of 5th phalange	Difficult to palpate	Ulnar nerve (C8, T1)
Abductor digiti minimi	Pisiform and flexor carpi ulnaris tendon	Ulnar aspect of base of 5th proximal phalanx	MCP abduction of 5th phalange	Ulnar border of 5th metacarpal	Ulnar nerve (C8, T1)
Flexor digiti minimi brevis	Hook of hamate and adjacent transverse carpal ligament	Ulnar aspect of base of 5th proximal phalanx	MCP flexion of 5th phalange	Palmar surface of 5th metacarpal	Ulnar nerve (C8, T1)

Web sites

Anatomy & Physiology Tutorials:

www.gwc.maricopa.edu/class/bio201/index.htm

Radiologic Anatomy Browser:

http://radlinux1.usuf1.usuhs.mil/rad/iong/index.html

This site has numerous radiological views of the musculoskeletal system.

University of Arkansas Medical School Gross Anatomy for Medical Students:

anatomy.uams.edu/htmlpages/anatomyhtml/gross.html

Dissections, anatomy tables, atlas images, links, etc.

Loyola University Medical Center: Structure of the Human Body:

www.meddean.luc.edu/lumen/MedEd/GrossAnatomy/GA.html

An excellent site with many slides, dissections, tutorials, etc., for the study of human anatomy.

Wheeless' Textbook of Orthopaedics:

www.medmedia.com/

This site has an extensive index of links to the fractures, joints, muscles, nerves, trauma, medications, medical topics, and lab tests, as well as links to orthopedic journals and other orthopedic and medical news.

The University of Texas MD Anderson Cancer Center Multimedia and Learning Resources:

rpiwww.mdacc.tmc.edu/mmlearn/anatomy.html

This site has numerous cadaveric cuts of the foot, knee, hand, arm, and elbow; an interactive ankle; and a rotating foot and ankle.

Arthroscopy.Com:

www.arthroscopy.com/sports.htm

Patient information on various musculoskeletal problems of the upper and lower extremity

Electronic Textbook of Hand Surgery:

www.e-hand.com/default.htm

Slides and illustrations of upper extremity musculoskeletal anatomy

Premiere Medical Search Engine:

www.medsite.com

This site allows the reader to enter any medical condition and it will search the net to find relevant articles.

Virtual Hospital:

www.vh.org

Numerous slides, patient information, etc.

Dynamic Human version 2.0 CD-ROM: The Visual Guide to Anatomy & Physiology:

www.mhhe.com/biosci/ap/dynamichuman2/

Web site that accompanies this CD-ROM

Dynamic Human CD activities

1. Review anatomical landmarks as well as origins/insertions for the radius, ulna and hand as they pertain to the joints of the wrist and hand. To do this, click on **skeletal**; then **anatomy, gross anatomy, upper limbs**, and **radius/ulna**; and then **hand**.
2. Review each of the muscles from this chapter by clicking on **muscular; anatomy, body regions**, and **forearm**.
3. Which muscles and what actions would be affected by carpal tunnel syndrome? Click on **nervous** and **clinical concepts**, then **carpal tunnel syndrome**.

Worksheet exercises

As an aid to learning, for in-class or out-of-class assignments, or for testing, tear-out worksheets are found at the end of the text (pp. 258 and 259).

Skeletal worksheet (no. 1)

Draw and label on the worksheet the following muscles:
a. Flexor pollicis longus
b. Flexor carpi radialis
c. Flexor carpi ulnaris
d. Extensor digitorum
e. Extensor pollicis longus
f. Extensor carpi ulnaris

Human figure worksheet (no. 2)

Label and indicate with arrows the following movements of the wrist and hand:
a. Flexion
b. Extension
c. Abduction (ulnar flexion)
d. Adduction (radial flexion)

Laboratory and review exercises

1. Locate the following parts of the humerus, radius, ulna, carpals, and metacarpals on a human skeleton and on a subject:
   a. Skeleton
      1. Medial epicondyle
      2. Lateral epicondyle
      3. Trochlea
      4. Capitulum
      5. Coronoid process
      6. Tuberosity of the radius
      7. Styloid process—radius
      8. Styloid process—ulna
      9. First and third metacarpals
      10. Wrist bones
      11. First phalanx of third metacarpal
   b. Subject
      1. Medial epicondyle
      2. Lateral epicondyle
      3. Pisiform
      4. Scaphoid (navicular)
2. How and where can the following muscles be palpated on a human subject?
   a. Flexor pollicis longus
   b. Flexor carpi radialis
   c. Flexor carpi ulnaris
   d. Extensor digitorum communis
   e. Extensor pollicis longus
   f. Extensor carpi ulnaris
3. Demonstrate the action and list the muscles primarily responsible for these movements at the wrist joint:
   a. Flexion
   b. Extension
   c. Abduction
   d. Adduction
4. List the planes in which each of the following wrist, hand, and finger joint movements occur. List the respective axis of rotation for each movement in each plane.
   a. Abduction
   b. Adduction
   c. Flexion
   d. Extension
5. Discuss why the thumb is the most important part of the hand.
6. How should boys and girls be taught to do push-ups? Justify your answer.
   a. Hands flat on the floor
   b. Fingertips
7. With a laboratory partner, determine how and why maintaining full flexion of all the fingers is impossible when passively moving the wrist into maximal flexion. Is it also difficult to maintain maximal extension of all the finger joints while passively taking the wrist into full extension?
8. Fill in the muscle analysis chart by listing the muscles primarily involved in each movement.

Muscle analysis chart • Wrist, hand, and fingers

Wrist and hand	
Flexion	Extension
Adduction	Abduction

Fingers—metacarpophalangeal joints	
Flexion	Extension

Fingers—proximal interphalangeal joints	
Flexion	Extension

Fingers—distal interphalangeal joints	
Flexion	Extension

Thumb	
Flexion	Extension

References

Gabbard CP, et al: Effects of grip and forearm position on flex arm hang performance, *Research Quarterly for Exercise and Sport,* July 1983.

Gench BE, Hinson MM, Harvey PT: *Anatomical kinesiology,* Dubuque, IA, 1995, Eddie Bowers.

Herrick RT, Herrick S: Ruptured triceps in powerlifter presenting as cubital tunnel syndrome—a case report, *American Journal of Sports Medicine* 15:514, September-October 1987.

Hislop HJ, Montgomery J: *Daniels and Worthingham's muscle testing: techniques of manual examination,* ed 6, Philadelphia, 1995, Saunders.

Lindsay DT: *Functional human anatomy,* St. Louis, 1996, Mosby.

Luttgens K, Hamilton N: *Kinesiology: scientific basis of human motion,* ed 9, Madison, WI, 1997, Brown & Benchmark.

Norkin CC, Levangie PK: *Joint structure and function—a comprehensive analysis,* Philadelphia, 1983, Davis.

Norkin CC, White DJ: *Measurement of joint motion: a guide to goniometry,* Philadelphia, 1985, Davis.

Rasch PJ: *Kinesiology and applied anatomy,* ed 7, Philadelphia, 1989, Lea & Febiger.

Seeley RR, Stephens TD, Tate P: *Anatomy & physiology,* ed 2, St. Louis, 1992, Mosby-Year Book.

Sieg KW, Adams SP: *Illustrated essentials of musculoskeletal anatomy,* ed 2, Gainesville, FL, 1985, Megabooks.

Sisto DJ, et al: An electromyographic analysis of the elbow in pitching, *American Journal of Sports Medicine* 15:260, May-June 1987.

Smith LK, Weiss EL, Lehmkuhl LD: *Brunnstrom's clinical kinesiology,* ed 5, Philadelphia, 1996, Davis.

Springer SI: Racquetball and elbow injuries, *National Racquetball* 16:7, March 1987.

Stone RJ, Stone JA: *Atlas of the skeletal muscles,* 1990, McGraw-Hill Companies, Inc., New York.

Muscular analysis of the upper extremity

6

Objectives

● To analyze the joint movements and muscles used in simple exercises

● To learn and understand the concept of open versus closed kinetic chain

● To learn to group individual muscles into units that produce certain joint movements

● To begin to think of exercises that increase the strength and endurance of individual muscle groups

The upper extremity is often one of the body's weakest areas when considering the number of muscles involved. American boys and girls are extremely weak in the upper shoulder area. A majority are unable to do one chin-up. In a traditional chin-up (pull-up), the subject grasps a horizontal bar with the palms toward the face and feet off the floor. The body is then pulled up until the chin is over the bar. In a modified chin-up (pull-up), the feet are on the floor; the subject then grasps the horizontal bar and pulls up the body to touch the bar. The traditional chin-up (pull-up) had to be modified to secure more meaningful results.*

Strength and endurance in this part of the human body are essential for improved appearance and posture, as well as for more efficient skill performance. Specific exercises and activities to condition this area should be intelligently selected by becoming thoroughly familiar with the muscles involved.

At this stage, simple exercises are used to begin teaching individuals how to group muscles to produce joint movement. Some of these simple introductory exercises are included in this chapter.

The early analysis of exercise makes the study of structural kinesiology more meaningful as students come to better understand the importance of individual muscles and groups of muscles in bringing about joint movements in various exercises. Chapter 11 contains analysis of exercises for the entire body, with emphasis on the trunk and lower extremities. Contrary to what most beginning students in structural kinesiology believe, muscular analysis of activities is not difficult once the basic concepts are understood.

---

*Pate R, et al: The national children and youth fitness study, II. The modified pull-up test, *Journal of Physical Education, Recreation and Dance* 58:71, 1987.

Concepts for analysis

In analyzing activities, it is important to understand that muscles are usually grouped according to their concentric function and work in paired opposition to an antagonistic group. An example of this *aggregate* muscle grouping to perform a given joint action is seen with the quadriceps all working to cause knee extension in opposition to the hamstring, which all work together to cause knee flexion. An often confusing aspect is that, depending on the activity, these muscle groups can function to control the exact opposite actions by contracting eccentrically. That is, through eccentric contractions, the quadriceps may control knee flexion, and the hamstrings may control knee extension. Students should be able to view an activity and to not only determine which muscles are performing the movement but also know what type of contraction is occurring and what kind of exercises are appropriate for developing the muscles. Chapter 1 provides a review of how muscles contract to work in groups to function in joint movement.

Analysis of movement

When analyzing various exercises and sport skills, it is essential to break down all of the movements into phases. The number of phases, usually three to five, will vary, depending on the skill. Practically all sport skills will have at least a preparatory phase, a movement phase, and a follow-through phase. Many will also begin with a stance phase and end with a recovery phase. The names of the phases will vary from skill to skill to fit in with the terminology used in various sports.

The *stance phase* allows the athlete to assume a comfortable and balanced body position from which to initiate the sport skill. The emphasis is on setting the various joint angles in the correct positions with respect to one another and to the sport surface.

The *preparatory phase*, often referred to as the cocking or wind-up phase, is used to lengthen the appropriate muscles so that they will be in position to generate more force and momentum as they concentrically contract in the next phase.

The *movement phase*, sometimes known as the acceleration, action, motion, or contact phase, is the action part of the skill. It is the phase in which the summation of force is generated directly to the ball, sport object, or opponent and is usually characterized by near-maximal concentric activity in the involved muscles.

The *follow-through phase* begins immediately after the climax of the movement phase, in order to bring about negative acceleration of the involved limb or body segment. In this phase, often referred to as the deceleration phase, the velocity of the body segment progressively decreases, usually over a wide range of motion. This velocity decrease is usually attributable to high eccentric activity in the muscles that were antagonist to the muscles used in the movement phase.

The *recovery phase* is used after follow-through to regain balance and positioning to be ready for the next sport demand.

Skill analysis can be seen with the example of a baseball pitch in Fig. 6.1. The stance phase begins when the player assumes a position with the ball in the glove before receiving the signal from the catcher. The pitcher begins the preparatory phase by extending the throwing arm posteriorly and rotating the trunk to the right in conjunction with left hip flexion. The right shoulder girdle is fully retracted in combination with abduction and maximum external rotation of the glenohumeral joint to complete this phase. Immediately

FIG. 6.1 ● Skill analysis phases—baseball pitch.

following, the movement phase begins with forward movement of the arm and continues until ball release. At ball release, the follow-through phase begins as the arm continues moving in the same direction established by the movement phase until the velocity decreases to the point that the arm can safely change movement direction. This deceleration of the body, and especially the arm, is accomplished by high amounts of eccentric activity. At this point, the recovery phase begins, enabling the player to reposition to field the batted ball. In this example, reference has been made briefly only to the throwing arm, but there are many similarities in other overhand sports skills such as the tennis serve, javelin throw, and volleyball serve. In actual practice, the movements of each joint in the body should be analyzed into the various phases.

Upper extremity activities

Children seem to have an innate desire to climb, swing, and hang. Such movements use the muscles of the hands, wrists, elbows, and shoulder joints. But the opportunity to perform these types of activities is limited in our modern culture. Unless emphasis is placed on the development of this area of our bodies by physical education teachers in elementary schools, for both boys and girls, it will continue to be muscularly the weakest area of our bodies. Weakness in the upper extremities can impair skill development and performance in many common enjoyable recreational activities, such as golf, tennis, softball, and racquetball. Athletes enjoy what they can do well, and they can be taught to enjoy activities that will increase the strength of this part of the body. These and other such activities are ones that can be enjoyed throughout the entire adult life; therefore, adequate skill development built on an appropriate base of muscular strength and endurance is essential for enjoyment and prevention of injury.

The kinetic chain concept

As you have learned, our extremities consist of several bony segments linked by a series of joints. These bony segments and their linkage system of joints may be likened to a chain. Just as with a chain, any one link in the extremity may be moved individually without significantly affecting the other links if the chain is open or not attached at one end. However, if the chain is securely attached or closed, substantial movement of any one link cannot occur without substantial and subsequent movement of the other links.

In the body, an extremity may be seen as representing an *open kinetic chain* if the distal end of the extremity is not fixed to any surface. This arrangement allows any one joint in the extremity to move or function separately without necessitating movement of other joints in the extremity. An example for the upper extremity is a shoulder shrug. If the distal end of the extremity is fixed, as in a push-up, the extremity represents a *closed kinetic chain*. In this closed system, movement of one joint cannot occur without causing predictable movements of the other joints in the extremity.

Consideration of the open versus closed kinetic chain through analysis of skilled movements is helpful in determining appropriate conditioning exercises to improve performance. Generally, closed kinetic chain exercises are more functional and applicable to the demands of sports and physical activity. Most sports involve closed-chain activities in the lower extremities and open-chain activities in the upper extremities. However, there are many exceptions, and closed-chain conditioning exercises may be beneficial for extremities primarily involved in open-chain sporting activities. Open-chain exercises generally isolate only one segment, while closed-chain exercises work all segments in the chain, resulting in conditioning of the muscles crossing each joint.

Analysis of upper body exercises

Presented over the next several pages are brief analyses of several common upper body exercises. Students are encouraged to follow and perhaps expand on the approach used to analyze other upper body activities. All muscles listed in the analysis are contracting concentrically unless specifically noted to be contracting eccentrically or isometrically.

Shoulder pull

Description

In a standing or sitting position, the subject interlocks the fingers in front of the chest and then attempts to pull them apart (Fig. 6.2). This contraction is maintained from 5 to 20 seconds.

Analysis

In this type of exercise, there is little or no movement of the contracting muscles. In certain isometric exercises, contraction of the antagonistic muscles is as strong as contraction of the muscles attempting to produce the force for movement. The muscle groups contracting to produce a movement are designated the *agonists*. In the exercise just described, the agonists in the right upper extremity are antagonistic to the agonists in the left upper extremity and vice versa. This exercise results in isometric contractions of the wrist and hand, elbow, shoulder joint, and shoulder girdle muscles. The strength of the contraction depends on the angle of pull and the leverage of the joint involved. Thus, it is not the same at each point.

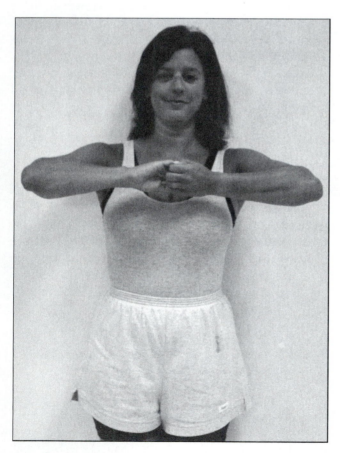

FIG. 6.2 ● Shoulder pull.

Attempted movements

NOTE: This entire exercise is isometric in design so that all contractions presented are isometric.

Extension of wrist and hand—resisted by flexors of wrists and hand
> Agonists—wrist and hand flexors
> Antagonists—wrist and hand extensors

Flexion of wrist and hand—resisted by extensors of wrist and hand
> Agonists—wrist and hand flexors
> Antagonists—wrist and hand extensors

Extension of elbow joint—resisted by flexors of wrist, elbow, and hand
> Agonists—triceps brachii and anconeus
> Antagonists—biceps brachii, brachialis, brachioradialis

Flexion of elbow joint—resisted by extensors of wrist, elbow, and hand
> Agonists—biceps brachii, brachialis, brachioradialis
> Antagonists—triceps brachii and anconeus

Abduction of shoulder joint—resisted by adductors of shoulder joint
> Agonists—deltoid and supraspinatus
> Antagonists—teres major, latissimus dorsi, pectoralis major

Adduction of shoulder joint—resisted by abductors of shoulder joint
> Agonists—teres major, latissimus dorsi, pectoralis major
> Antagonists—deltoid and supraspinatus

Adduction and depression of shoulder girdle—resisted by abductors
> Agonists—rhomboid and trapezius
> Antagonists—serratus anterior, pectoralis minor, trapezius (upper)

Abduction and elevation of shoulder girdle—resisted by adductors
> Agonists—serratus anterior, pectoralis minor, trapezius (upper)
> Antagonists—rhomboid and trapezius

Isometric exercises vary in the number of muscles contracting, depending on the type of exercise and the joints at which movement is attempted. The shoulder pull exercise produces some contraction of antagonistic muscles at four sets of joints.

Arm curl

Description

With the subject in a standing position, the dumbbell is held in the hands with the palms to the front. The dumbbell is lifted until the elbow is completely flexed (Fig. 6.3). Then it is returned to the starting position.

Analysis

This exercise is divided into two movements for analysis: (1) upward curl movement and (2) return movement to starting position. NOTE: An assumption is made that no movement occurs in the shoulder joint and shoulder girdle.

Upward curl movement
    Wrist and hand. NOTE: The wrist is in a position of slight extension to facilitate greater active finger flexion in gripping the dumbbell. (The flexors remain isometrically contracted throughout the entire exercise, to hold the dumbbell.)

        Flexion
           Wrist and hand flexors
              Flexor carpi radialis
              Flexor carpi ulnaris
              Palmaris longus
              Flexor digitorum profundus
              Flexor digitorum superficialis
              Flexor pollicis longus
    Elbow joint
        Flexion
           Elbow flexors
              Biceps brachii
              Brachialis
              Brachioradialis
Return movement to starting position
    Wrist and hand
        Flexion
           Wrist and hand flexors (isometric contraction)
    Elbow joint
        Extension
           Elbow flexors (eccentric contraction)

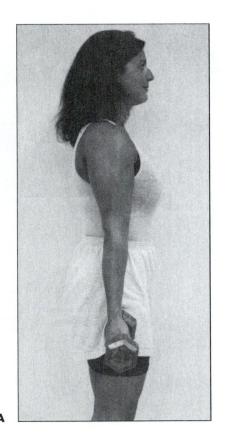 

A   B

FIG. 6.3 ● Arm curl. **A**, Beginning position in extension; **B**, flexed position.

Triceps extension

Description

The subject uses the opposite hand to assist in maintaining the arm in a shoulder flexed position. The subject then, grasping the dumbbell and beginning in full elbow flexion, extends the elbow until the arm and forearm are straight. The shoulder joint and shoulder girdle are stabilized by the opposite hand. Consequently, no movement is assumed to occur in these areas (Fig. 6.4).

Analysis

This exercise can be divided into two movements for analysis: (1) movement of pressing arms down to straight-arm position and (2) return movement to starting position.

Movement to straight-arm position
    Wrist and hand. NOTE: The wrist is in a position of slight extension to facilitate greater active finger flexion in gripping the bar. (The flexors remain isometrically contracted throughout the entire exercise, to hold the dumbbell)

    Flexion
        Wrist and hand flexors
            Flexor carpi radialis
            Flexor carpi ulnaris
            Palmaris longus
            Flexor digitorum profundus
            Flexor digitorum superficialis
            Flexor pollicis longus
    Elbow joint
        Extension
            Elbow extensors
                Triceps brachii
                Anconeus

Return movement to starting position
    Wrist and hand
        Flexion
            Wrist and hand flexors (isometric contraction)
    Elbow joint
        Flexion
            Elbow extensors (eccentric contraction)

FIG. 6.4 ● Triceps extension.

Barbell press

Description

This exercise is sometimes referred to as the *overhead* or *military press*. The barbell is held in a position high in front of the chest, with palms facing forward, feet comfortably spread, and back and legs straight (Fig. 6.5, *A*). From this position, the barbell is pushed upward, until fully overhead, and then it is returned to the starting position (Fig. 6.5, *B*).

Analysis

This exercise is separated into two movements for analysis: (1) upward movement and (2) return movement to starting position.

Upward movement
    Wrist and hand. NOTE: The wrist is in a position of slight extension to facilitate greater active finger flexion in gripping the bar.
        No movement
            Wrist and hand flexors (isometric contraction)
    Elbow joint
        Extension
            Elbow extensors
                Triceps brachii
                Anconeus
    Shoulder joint
        Flexion
            Shoulder joint flexors
                Pectoralis major (clavicular head or upper fibers)
                Anterior deltoid
                Coracobrachialis
                Biceps brachii
    Shoulder girdle
        Upward rotation and elevation
            Shoulder girdle upward rotators and elevators
                Trapezius
                Levator scapulae
                Serratus anterior
Return movement to starting position
    Wrist and hand
        No movement
            Wrist and hand flexors (isometric contraction)
    Elbow joint
         Flexion
             Elbow joint extensors (eccentric contraction)
    Shoulder joint
        Extension
            Shoulder joint flexors (eccentric contraction)
    Shoulder girdle
        Downward rotation and depression
            Shoulder girdle upward rotators and elevators (eccentric contraction)

A

B

FIG. 6.5 ● Barbell press. **A**, Starting position; **B**, full overhead position.

Chest press (bench press) FIG. 6.6

Description

The subject lies on the exercise bench in the supine position, grasps the barbell, and presses the weight upward through the full range of arm and shoulder movement. Then the weight is lowered to the starting position.

Analysis

The chest press can be divided into two movements for analysis: (1) upward movement to extend the elbows and (2) return movement to the starting position.

Movement to upward position
    Wrist and hand. NOTE: The wrist is in a position of slight extension to facilitate greater active finger flexion in gripping the bar.
        Flexion
            Wrist and hand flexors (isometric contraction)
                Flexor carpi radialis
                Flexor carpi ulnaris
                Palmaris longus
                Flexor digitorum profundus
                Flexor digitorum superficialis
                Flexor pollicis longus
    Elbow joint
        Extension
            Elbow extensors
                Triceps brachii
                Anconeus

Shoulder joint
    Flexion and horizontal adduction
        Shoulder joint flexors and horizontal adductors
            Pectoralis major
            Anterior deltoid
            Coracobrachialis
            Biceps brachii
Shoulder girdle
    Abduction
        Shoulder girdle abductors
            Serratus anterior
            Pectoralis minor
Return movement to starting position
    Wrist and hand
        Flexion
            Wrist and hand flexors (eccentric contraction)
    Elbow joint
        Flexion
            Elbow extensors (eccentric contraction)
    Shoulder joint
        Extension and horizontal abduction
            Shoulder joint flexors and horizontal adductors (eccentric contraction)
    Shoulder girdle
        Adduction
            Shoulder girdle abductors (eccentric contraction)

A

B

FIG. 6.6 ● Chest press (bench press). **A,** Starting position; **B,** up position.

Chin-up (pull-up)

Description

The subject grasps a horizontal bar or ladder with the palms toward the face (Fig. 6.7, *A*). From a hanging position on the bar, the subject pulls up until the chin is over the bar (Fig. 6.7, *B*) and returns to the starting position (Fig. 6.7, *C*).

Analysis

This exercise is separated into two movements for analysis: (1) movement upward to chinning position (concentric phase) and (2) return movement to hanging position (eccentric phase).

Movement upward to chinning position
    Wrist and hand
        Isometric contraction of wrist and hand
           flexors
    Elbow joint
        Flexion
           Elbow flexors
               Biceps brachii
               Brachialis
               Brachioradialis
    Shoulder joint
        Extension
           Shoulder joint extensors
               Latissimus dorsi
               Teres major
               Posterior deltoid
               Pectoralis major
               Triceps brachii (long head)
    Shoulder girdle
        Adduction, depression, and downward
          rotation
           Shoulder girdle adductors, depressors,
             and downward rotators
               Trapezius (lower and middle)
               Pectoralis minor
               Rhomboids
Return movement to hanging position
    Wrist and hand
        Flexion
           Wrist and hand flexors
    Elbow joint
        Extension
           Elbow joint flexors (eccentric
             contraction)
    Shoulder joint
        Flexion
           Shoulder joint extensors (eccentric
             contraction)
    Shoulder girdle
        Elevation, abduction, and upward rotation
           Shoulder girdle adductors, depressors,
             and downward rotators

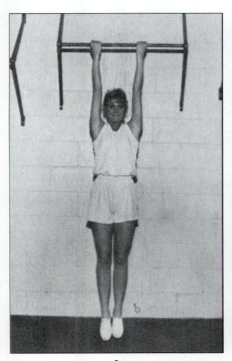

A   **B**   **C**

FIG. 6.7 ● Pull-up. **A**, Straight-arm hang; **B**, chin over bar; **C**, bent-arm hang on way up or down.

Push-up (fingertip)

Description

The subject lies on the floor in a prone position with the legs together, the fingertips touching the floor, and the hands pointed forward and approximately under the shoulders (Fig. 6.8, *A*). Keeping the back and legs straight, the subject pushes up to the up position and returns to the starting position (Fig. 6.8, *B*).

Analysis

This exercise is separated into two movements for analysis: (1) movement to the up position (concentric phase) and (2) return movement to the starting position (eccentric phase).

Movement to up position
    Wrist and hand
        Isometric contraction of wrist and hand
           flexors and extensors
    Elbow joint
        Extension
           Elbow extensors
               Triceps brachii
               Anconeus
    Shoulder joint
        Horizontal adduction
           Shoulder joint horizontal adductors
               Pectoralis major
               Anterior deltoid
               Biceps brachii
               Coracobrachialis
    Shoulder girdle
        Abduction
           Shoulder girdle abductors
               Serratus anterior
               Pectoralis minor
Return movement to starting position
    Wrist and hand
        Isometric contraction of wrist and hand
           flexors and extensors
    Elbow joint
        Flexion
           Elbow joint extensors (eccentric
               contraction)
    Shoulder joint
        Horizontal abduction
           Shoulder joint horizontal adductors
               (eccentric contraction)
    Shoulder girdle
        Adduction
           Shoulder girdle abductors (eccentric
               contraction)

Chin-ups and push-ups are excellent exercises for the shoulder area, shoulder girdle, shoulder joint, elbow joint, and wrist and hand (see Figs. 6.7 and 6.8). The use of free weights, machines, and other conditioning exercises helps develop strength and endurance for this part of the body.

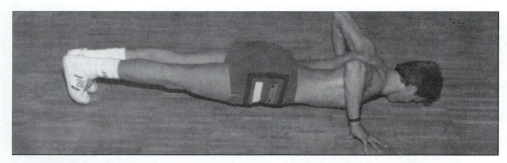

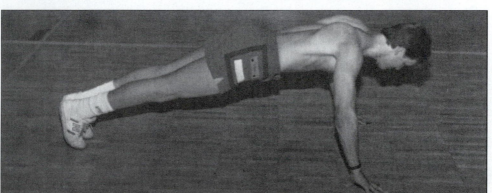

FIG. 6.8 ● Push-up. **A**, Starting position; **B**, up position.

Latissimus pull (lat pull)

Description

From a sitting position, the subject reaches up and grasps a horizontal bar (Fig. 6.9, *A*). The bar is pulled down to a position behind the neck and shoulders (Fig. 6.9, *B*). Then it is returned slowly to the starting position.

Analysis

This exercise is separated into two movements for analysis: (1) movement downward to a position behind the shoulders (concentric phase) and (2) return to the starting position (eccentric phase).

Movement downward to position behind shoulders
    Wrist and hand
        Isometric contraction of wrist and hand
          flexors
    Elbow joint
        Flexion
            Elbow flexors
                Biceps brachii
                Brachialis
                Brachioradialis
    Shoulder joint
        Adduction
            Shoulder joint adductors
                Pectoralis major
                Posterior deltoid
                Latissimus dorsi
                Teres major
                Subscapularis
    Shoulder girdle
        Adduction, depression, and downward
          rotation
            Shoulder girdle adductors, depressors,
              and downward rotators
                Trapezius (lower)
                Rhomboid
                Pectoralis minor
Return movement to starting position
    Wrist and hand
        Isometric contraction of wrist and hand
          flexors
    Elbow joint
        Extension
            Elbow joint flexors (eccentric
              contraction)
    Shoulder joint
        Abduction
            Shoulder joint adductors (eccentric
              contraction)
    Shoulder girdle
        Abduction, elevation, and upward rotation
            Shoulder girdle adductors, depressors,
              and downward rotators (eccentric
              contraction)

 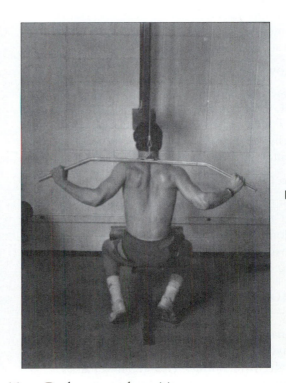

FIG. 6.9 ● Latissimus pull (lat pull). **A,** Starting position; **B,** downward position.

Web sites

Stretching and Flexibility: Everything you never wanted to know:

www.cs.huji.ac.il/papers/rma/stretching_toc.html

This paper by Brad Appleton gives detailed information on stretching and stretching techniques. It includes normal ranges of motion, flexibility, how to stretch, the physiology of stretching, and the types of stretching, including PNF.

Fitness World:

www.fitnessworld.com

The information at this site is about fitness in general and includes access to *Fitness Management* magazine.

Concept II:

www.concept2.com/index.html

Information on the technique of rowing and the muscles used

Pump It Up!:

www.netspace.org/~gch/cs92/home.html

A site that has stretching and strengthening exercises for the major muscle groups of the body

Worksheet exercises

As an aid to learning, for in-class and out-of-class assignments, or for testing, a tear-out worksheet is found at the end of the text (p. 260).

Upright row exercise worksheet (no. 1)

List the movements that occur in each joint as the subject lifts the weight in performing upright rows. For each joint movement, list the muscles primarily responsible and denote whether they are contracting concentrically or eccentrically.

Dips exercise worksheet (no. 2)

List the movements that occur in each joint as the subject moves the body in performing dips. For each joint movement, list the muscles primarily responsible and denote whether they are contracting concentrically or eccentrically.

Laboratory and review exercises

1. Analyze other conditioning exercises that involve the shoulder areas, such as dips, upright rows, shrugs, and inclined presses.
2. Observe and analyze shoulder muscular activities of children on playground equipment.
3. Discuss how you would teach boys and girls who cannot perform one chin-up to learn to do chin-ups. Additionally, discuss how you would teach subjects who cannot perform one push-up to do push-ups.
4. Should boys and girls attempt to do chin-ups and push-ups to see whether they have adequate strength in the shoulder area?
5. Test yourself doing chin-ups and push-ups to determine your strength and muscular endurance in the shoulder area.
6. What, if any, benefit would result from doing fingertip push-ups as opposed to push-ups with the hands flat on the floor?
7. Stand slightly farther than arm's length from a wall with your hands facing forward at shoulder-level height. Push your hand straight in front of your shoulder until the elbow is extended fully and the glenohumeral joint is flexed 90 degrees by performing each of the following movements before proceeding to the next:
   • Glenohumeral flexion to 90 degrees
   • Full elbow extension
   • Wrist extension to 70 degrees

   Analyze the movements and the muscles responsible for each movement at the shoulder girdle, glenohumeral joint, elbow, and wrist.
8. Now face a wall and stand about 6 inches from it. Place both hands on the wall at shoulder level and put your nose and chest against the wall. Keeping your palms in place on the wall, slowly push your body from the wall as in a push-up until your chest is as far away from the wall as possible without removing your palms from the wall surface. Analyze the movements and the muscles responsible for each movement at the shoulder girdle, glenohumeral joint, elbow, and wrist.
9. What is the difference between the two exercises in questions #7 and #8? Can you perform #8 one step at a time, as you did in #7?
10. Analyze each exercise in the exercise analysis chart. Use one row for each joint involved that actively moves during the exercise. Do not include joints where there is no active movement or where the joint is maintained in one position isometrically.

Exercise analysis chart

Exercise	Phase	Joint, movement occurring	Force causing movement (muscle or gravity)	Force resisting movement (muscle or gravity)	Functional muscle group, type of contraction
Barbell press (overhead or military press)	Upward movement				
	Return movement				
Chest press (bench press)	Upward movement				
	Return movement				
Chin-up (pull-up)	Upward movement				
	Return movement				
Push-up	Upward movement				
	Return movement				
Latissimus pull (lat pull)	Downward movement behind shoulders				
	Return movement				

References

Adrian M: Isokinetic exercise, *Training and Conditioning* 1:1, June 1991.

Andrews JR, Wilk KE: *The athlete's shoulder,* New York, 1994, Churchill Livingstone.

Andrews JR, Zarins B, Wilk KE: *Injuries in baseball,* Philadelphia, 1998, Lippincott-Raven.

Booher JM, Thibodeau GA: *Athletic injury assessment,* ed 4, 2000, McGraw-Hill Companies, Inc., New York.

Bouche J: Three essential lifts for high school players, *Scholastic Coach* 56:42, April 1987.

Brzycki M: Rx for a safe productive strength program, *Scholastic Coach* 57:70, September 1987.

Epley B: Getting elementary muscles, *Coach and Athlete* 44:60, November-December 1981.

Geisler P: Kinesiology of the full golf swing—implications for intervention and rehabilitation, *Sports Medicine Update* 11:9, #2, 1996.

Luttgens K, Hamilton N: *Kinesiology: scientific basis of human motion,* ed 9, Madison, WI, 1997, Brown & Benchmark.

Matheson O, et al: Stress fractures in athletes, *American Journal of Sports Medicine* 15:46, January-February 1987.

Northrip JW, Logan GA, McKinney WC: *Analysis of sport motion: anatomic and biomechanic perpectives,* ed 3, 1983, McGraw-Hill Companies, Inc., New York.

Schlitz J: The athlete's daily dozen stretches, *Athletic Journal* 66:20, November 1985.

Smith LK, Weiss EL, Lehmkuhl LD: *Brunnstrom's clinical kinesiology,* ed 5, Philadelphia, 1996, Davis.

Steindler A: *Kinesiology of the human body,* Springfield, IL, 1970, Charles C Thomas.

The hip joint and pelvic girdle

7 · · · · · · · · · · · · · · ·

Objectives

● To identify on a human skeleton or subject selected bony features of the hip joint and pelvic girdle

● To label on a skeletal chart selected bony features of the hip joint and pelvic girdle

● To draw on a skeletal chart the individual muscles of the hip joint

● To demonstrate, using a human subject, all of the movements of the hip joint and pelvic girdle and list their respective planes of movement and axes of motion

● To palpate on a human subject the muscles of the hip joint and pelvic girdle

● To list and organize the primary muscles that produce movement of the hip joint and pelvic girdle and list their antagonists

The hip, or acetabular femoral joint, is a relatively stable joint due to its bony architecture, strong ligaments, and large, supportive muscles. It functions in weight bearing and locomotion, which is enhanced significantly by its wide range of motion, which provides the ability to run, cross-over cut, side-step cut, jump, and make many other directional changes.

Bones   FIGS. 7.1 to 7.3

The hip joint is the ball-and-socket joint that consists of the head of the femur connecting with the acetabulum of the pelvic girdle. The pelvic girdle consists of a right and left pelvic bone joined together posteriorly by the sacrum. The femur is the longest bone in the body. The sacrum can be considered an extension of the spinal column with five fused vertebrae. Extending inferior to the sacrum is the coccyx. The pelvic bones are made up of three bones: the ilium, the ischium, and the pubis. At birth and during growth and development, they are three distinct bones. At maturity, they are fused to form one pelvic bone.

The pelvic bone can be divided roughly into three areas, starting from the acetabulum:

Upper two-fifths = ilium
Posterior and lower two-fifths = ischium
Anterior and lower one-fifth = pubis

Joints   FIGS. 7.1 to 7.3

In the anterior area, the pelvic bones are joined to form the symphysis pubis, an amphiarthrodial joint. In the posterior area, the sacrum is located between the two pelvic bones and forms the sacroiliac joints. Strong ligaments unite these bones to form rigid, slightly movable joints. The bones are large and heavy and for the most part are covered by thick, heavy muscles. Very minimal oscillating-type movements can occur in these joints, as in walking or in hip flexion when lying on one's back. However, movements usually involve the entire pelvic girdle and hip joints. In walking, there is hip flexion and extension with rotation of the pelvic girdle, forward in hip flexion and backward in hip extension. Jogging and running result in faster movements and in a greater range of movement.

Sport skills, such as kicking a football or soccer ball, are other good examples of hip and pelvic

FIG. 7.1 ● Right pelvis and femur, anterior view.

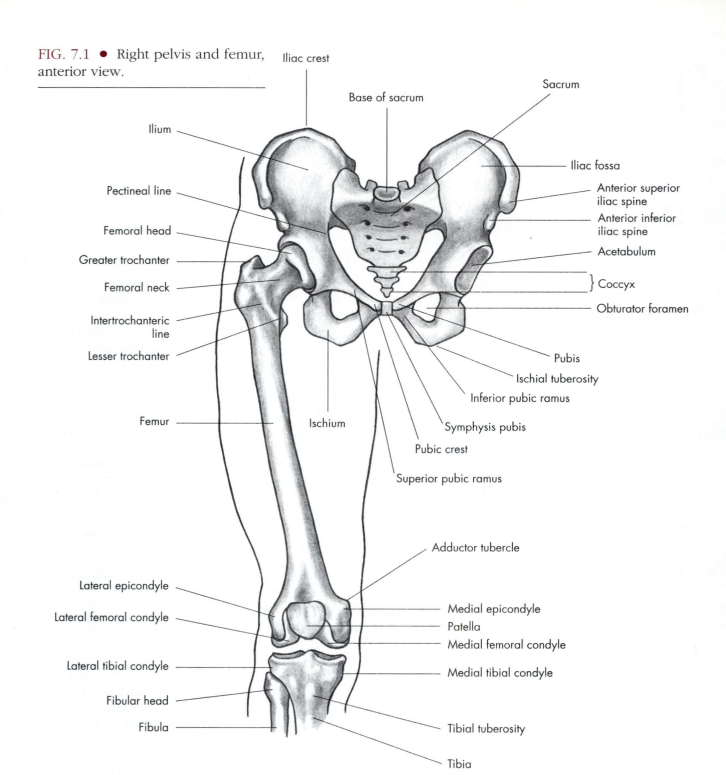

Iliac crest

Base of sacrum

Sacrum

Ilium

Iliac fossa

Pectineal line

Anterior superior iliac spine

Femoral head

Anterior inferior iliac spine

Greater trochanter

Acetabulum

Femoral neck

} Coccyx

Intertrochanteric line

Obturator foramen

Lesser trochanter

Pubis

Ischial tuberosity

Inferior pubic ramus

Femur

Ischium

Symphysis pubis

Pubic crest

Superior pubic ramus

Adductor tubercle

Lateral epicondyle

Medial epicondyle

Lateral femoral condyle

Patella

Medial femoral condyle

Lateral tibial condyle

Medial tibial condyle

Fibular head

Fibula

Tibial tuberosity

Tibia

movements. Pelvic rotation helps increase the length of the stride in running; in kicking, it can result in a greater range of motion, which translates to a greater distance or more speed to the kick.

Except for the glenohumeral joint, the hip is one of the most mobile joints of the body, largely because of its multiaxial arrangement. Unlike the glenohumeral, the hip joint's bony architecture provides a great deal of stability, resulting in relatively few hip joint subluxations and dislocations.

The hip joint is classified as an enarthrodial-type joint and is formed by the femoral head inserting into the socket provided by the acetabulum of the pelvis. An extremely strong and dense ligamentous capsule reinforces the joint, especially anteriorly. Anteriorly, the iliofemoral, or Y, ligament prevents hip hyperextension. The teres ligament attaches from deep in the acetabulum to a depression in the head of the femur and slightly limits adduction. The pubofemoral ligament is located

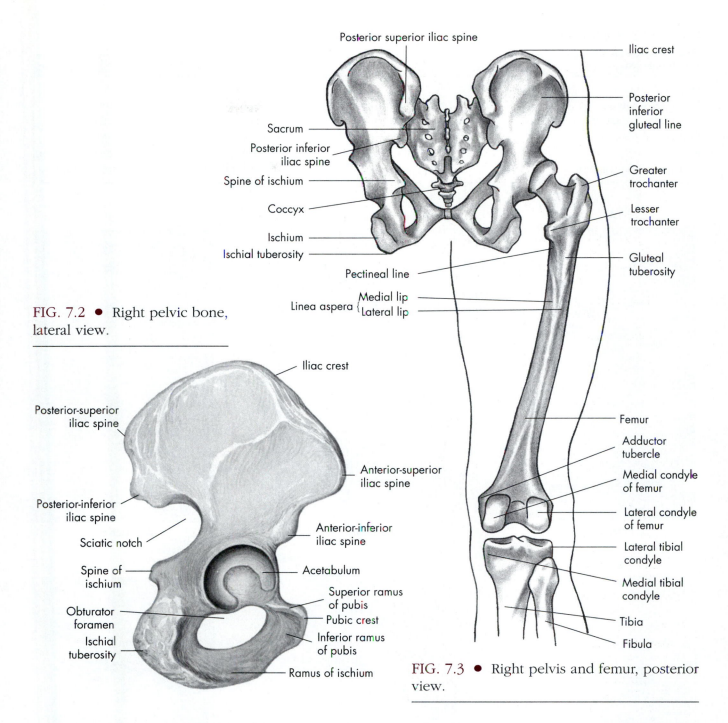

Posterior superior iliac spine

Iliac crest

Sacrum

Posterior inferior iliac spine

Spine of ischium

Coccyx

Ischium

Ischial tuberosity

Pectineal line

Medial lip
Linea aspera { Lateral lip

Posterior inferior gluteal line

Greater trochanter

Lesser trochanter

Gluteal tuberosity

FIG. 7.2 ● Right pelvic bone, lateral view.

Iliac crest

Posterior-superior iliac spine

Posterior-inferior iliac spine

Sciatic notch

Spine of ischium

Obturator foramen

Ischial tuberosity

Anterior-superior iliac spine

Anterior-inferior iliac spine

Acetabulum

Superior ramus of pubis

Pubic crest

Inferior ramus of pubis

Ramus of ischium

Femur

Adductor tubercle

Medial condyle of femur

Lateral condyle of femur

Lateral tibial condyle

Medial tibial condyle

Tibia

Fibula

FIG. 7.3 ● Right pelvis and femur, posterior view.

anteromedially and inferiorly and limits excessive extension and abduction. Posteriorly, the triangular ischiofemoral ligament extends from the ischium below to the trochanteric fossa of the femur and limits internal rotation.

Because of individual differences, there is some disagreement about the exact possible range of each movement in the hip joint, but the ranges are generally 0 to 130 degrees of flexion, 0 to 30 degrees of extension, 0 to 35 degrees of abduction, 0 to 30 degrees of adduction, 0 to 45 degrees of internal rotation, and 0 to 50 degrees of external rotation.

The pelvic girdle moves back and forth within three planes for a total of six different movements. To avoid confusion, it is important to analyze the pelvic girdle activity to determine the exact location of the movement. All pelvic girdle rotation actually results from motion at one or more of the following locations: the right hip, the left hip, the lumbar spine. Although it is not essential for movement to occur in all three of these areas, it must occur in at least one for the pelvis to rotate in any direction. Table 7.1 lists the motions at the hips and lumbar spine that can often accompany rotation of the pelvic girdle.

TABLE 7.1 • Motions accompanying pelvic rotation

Pelvic rotation	Lumbar spine motion	Right hip motion	Left hip motion
Anterior rotation	Extension	Flexion	Flexion
Posterior rotation	Flexion	Extension	Extension
Right lateral rotation	Left lateral flexion	Abduction	Adduction
Left lateral rotation	Right lateral flexion	Adduction	Abduction
Right transverse rotation	Left lateral rotation	Internal rotation	External rotation
Left transverse rotation	Right lateral rotation	External rotation	Internal rotation

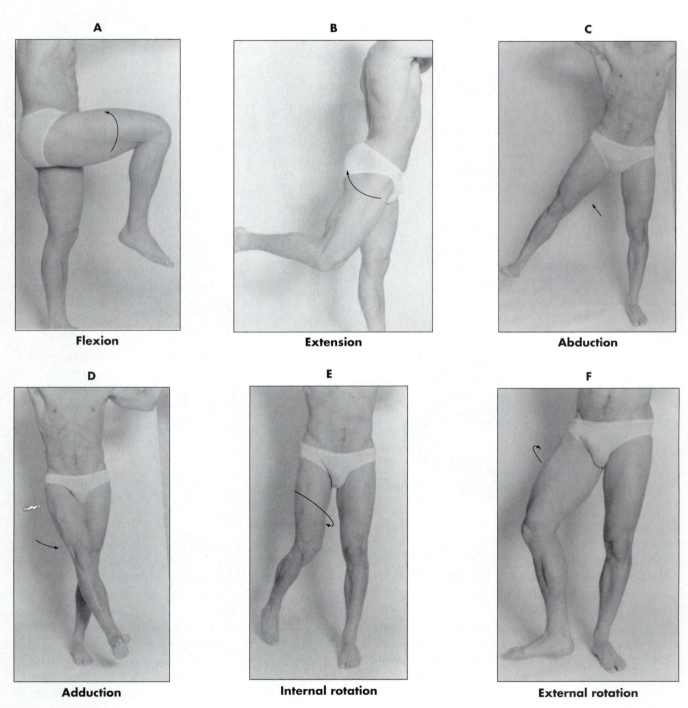

A

Flexion

B

Extension

C

Abduction

D

Adduction

E

Internal rotation

F

External rotation

FIG. 7.4 • Movements of the hip.

Movements   FIGS. 7.4 and 7.5

Anterior and posterior pelvic rotation occur in the sagittal or anteroposterior plane, whereas right and left lateral rotation occur in the lateral or frontal plane. Right transverse (clockwise) rotation and left transverse (counterclockwise) rotation occur in the horizontal or transverse plane of motion.

Hip flexion: movement of the femur straight anteriorly toward the pelvis

Hip extension: movement of the femur straight posteriorly away from the pelvis; sometimes referred to as hyperextension

Hip abduction: movement of the femur laterally to the side away from the midline

Hip adduction: movement of the femur medially toward the midline

Hip external rotation: rotary movement of the femur laterally around its longitudinal axis away from the midline; lateral rotation

Hip internal rotation: rotary movement of the femur medially around its longitudinal axis toward the midline; medial rotation

Hip diagonal abduction: movement of the femur in a diagonal plane away from the midline of the body

Hip diagonal adduction: movement of the femur in a diagonal plane toward the midline of the body

Anterior pelvic rotation: anterior movement of the upper pelvis; the iliac crest tilts forward in a sagittal plane; anterior tilt

Posterior pelvic rotation: posterior movement of the upper pelvis; the iliac crest tilts backward in a sagittal plane; posterior tilt

Left lateral pelvic rotation: in the frontal plane, the left pelvis moves inferiorly in relation to the right pelvis; either the left pelvis rotates downward or the right pelvis rotates upward; left lateral tilt

Right lateral pelvic rotation: in the frontal plane, the right pelvis moves inferiorly in relation to the left pelvis; either the right pelvis rotates downward or the left pelvis rotates upward; right lateral tilt

Left transverse pelvic rotation: in a horizontal plane of motion, the pelvis rotates to the body's left; the right iliac crest moves anteriorly in relation to the left iliac crest, which moves posteriorly

Right transverse pelvic rotation: in a horizontal plane of motion, the pelvis rotates to the body's right; the left iliac crest moves anteriorly in relation to the right iliac crest, which moves posteriorly

A	B	C	D

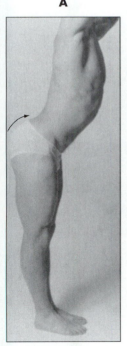 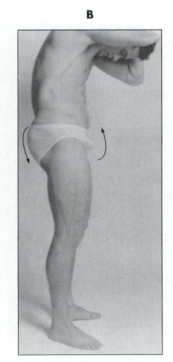 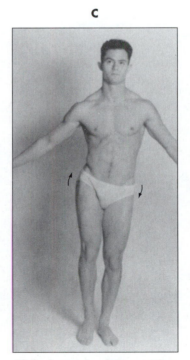 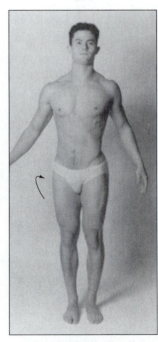

Anterior pelvic rotation   **Poster pelvic rotation**   **Right lateral pelvic rotation**   **Right transverse pelvic rotation**

FIG. 7.5 ● Pelvic girdle motions.

Muscles

At the hip joint, there are seven two-joint muscles that have one action at the hip and another at the knee. The muscles actually involved in hip and pelvic girdle motions depend largely on the direction of the movement and the position of the body in relation to the earth and its gravitational forces. In addition, it should be noted that the body part that moves the most will be the part least stabilized. For example, when standing on both feet and contracting the hip flexors, the trunk and pelvis will rotate anteriorly, but, when lying supine and contracting the hip flexors, the thighs will move forward into flexion on the stable pelvis.

In another example, the hip flexor muscles are used in moving the thighs toward the trunk, but the extensor muscles are used eccentrically when the pelvis and the trunk move downward slowly on the femur and concentrically when the trunk is raised on the femur—this, of course, in rising to the standing position.

In the downward phase of the knee-bend exercise, the movement at the hips and knees is flexion. The muscles primarily involved are the hip and knee extensors in eccentric contraction.

Hip joint and pelvic girdle muscles—location

Muscle location largely determines the muscle action. Seventeen or more muscles are found in the area (the six external rotators are counted as one muscle). Most hip joint and pelvic girdle muscles are large and strong.

Anterior
Primarily hip flexion
Iliopsoas
Pectineus
Rectus femoris*[†]
Sartorius[†]
Lateral
Primarily hip abduction
Gluteus medius
Gluteus minimus
External rotators
Tensor fasciae latae[†]
Posterior
Primarily hip extension
Gluteus maximus
Biceps femoris*[†]
Semitendinosus*[†]
Semimembranosus*[†]
External rotators (sixdeep)
Medial
Primarily hip adduction
Adductor brevis
Adductor longus
Adductor magnus
Gracilis[†]

---

*Two-joint muscles, knee actions are discussed in Chapter 8.
[†]Two-joint muscles.

The muscles of the pelvis that act on the hip joint may be divided into two regions—the iliac and gluteal regions. The iliac region contains the iliopsoas muscle, which flexes the hip. The iliopsoas actually is three different muscles—the iliacus, the psoas major, and the psoas minor. The 10 muscles of the gluteal region function primarily to extend and rotate the hip. Located in the gluteal region are the gluteus maximus, gluteus medius, gluteus minimi, tensor fasciae latae, and the six deep external rotators—piriformis, obturator externus, obturator internus, gemellus superior, gemellus inferior, and quadratus femoris.

The thigh is divided into three compartments by the intermuscular septa (Fig. 7.6). The anterior compartment contains the rectus femoris, vastus medialis, vastus intermedius, vastus lateralis, and sartorius. The hamstring muscle group, consisting of the biceps femoris, semitendinosus, and semimembranosus, is located in the posterior compartment. The medial compartment contains the thigh muscles primarily responsible for adduction of the hip, which are the adductor brevis, adductor longus, adductor magnus, pectineus, and gracilis.

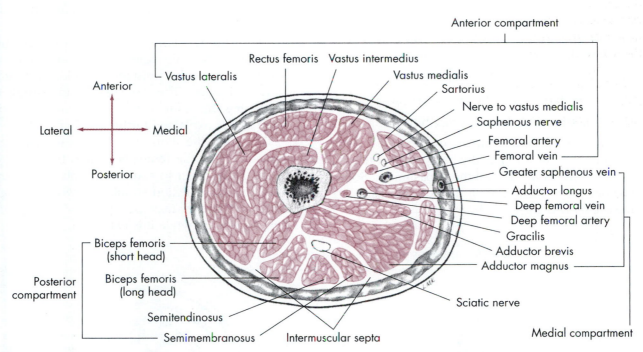

FIG. 7.6 ● Transverse section of the midthigh, detailing the anterior, posterior, and medial compartments.

Iliopsoas muscle   FIG. 7.7

(il´e-o-so´as)

Origin

Iliacus: inner surface of the ilium

Psoas major and minor: lower borders of the transverse processes (L1–5), sides of the bodies of the last thoracic vertebrae (T12), the lumbar vertebrae (L1–5), intervertebral fibrocartilages, and base of the sacrum

Insertion

Iliacus and psoas major: lesser trochanter of the femur and the shaft just below

Psoas minor: pectineal line and iliopectineal eminence

Action

Flexion of the hip

External rotation of the femur

Palpation

Impossible to palpate except with almost complete relaxation of the rectus abdominis muscle

Innervation

Lumbar nerve and femoral nerve (L2–4)

Application, strengthening, and flexibility

The iliopsoas is commonly referred to as if it were one muscle, but it is actually composed of the iliacus, the psoas major, and the psoas minor. Some anatomy texts make this distinction and list each muscle individually.

The iliopsoas muscle is powerful in actions such as raising the legs from the floor while in a supine position. The psoas major's origin in the lower back tends to move the lower back anteriorly or, in the supine position, pulls the lower back up as it raises the legs. For this reason, lower back problems are often aggravated by this activity and bilateral 6-inch leg raises are usually not recommended. The abdominals are the muscles that can be used to prevent this lower back strain by pulling up on the front of the pelvis, thus flattening the back. Leg raising is primarily hip flexion and not abdominal action. Backs may be injured by strenuous and prolonged leg-raising exercises. The iliopsoas contracts strongly, both concentrically and eccentrically, in sit-ups, particularly if the hip is not flexed.

The iliopsoas may be exercised by supporting the arms on a dip bar or parallel bars and then flexing the hips to lift the legs. This may be done initially with the knees flexed in a tucked position to lessen the resistance. As the muscle becomes more developed, the knees can be straightened, which increases the resistance arm length to add more resistance. This concept of increasing or decreasing the resistance by modifying the resistance arm is explained further in Chapter 12.

To stretch the iliopsoas, the hip must be extended so that the femur is behind the plane of the body. In order to somewhat isolate the iliopsoas, full knee flexion should be avoided. Slight additional stretch may be applied by internally rotating the hip while it is extended.

FIG. 7.7 ● Iliopsoas muscle, anterior view. *O*, Origin; *I*, insertion.

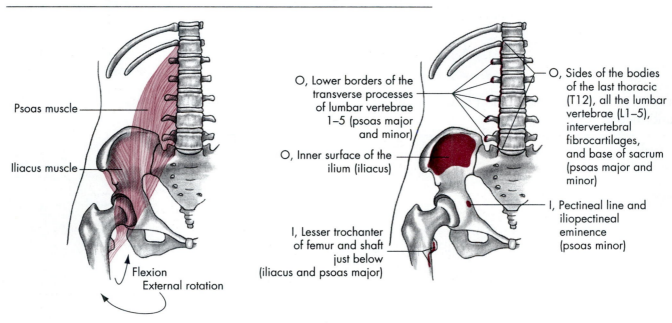

Psoas muscle

Iliacus muscle

Flexion
External rotation

O, Lower borders of the transverse processes of lumbar vertebrae 1–5 (psoas major and minor)

O, Inner surface of the ilium (iliacus)

I, Lesser trochanter of femur and shaft just below (iliacus and psoas major)

O, Sides of the bodies of the last thoracic (T12), all the lumbar vertebrae (L1–5), intervertebral fibrocartilages, and base of sacrum (psoas major and minor)

I, Pectineal line and iliopectineal eminence (psoas minor)

Sartorius muscle   FIG. 7.8

(sar-to´ri-us)

Origin

Anterior superior iliac spine and notch just below
  the spine

Insertion

Anterior medial condyle of the tibia

Action

Flexion of the hip
Flexion of the knee
External rotation of the thigh as it flexes the hip and
  knee

Palpation

Easiest to palpate at the anterior superior spine of
  the ilium; impossible to palpate on subjects with
  medium and heavy legs

Innervation

Femoral nerve (L2–3)

Application, strengthening, and flexibility

Pulling from the anterior superior iliac spine and
the notch just below it, the tendency again is to tilt
the pelvis anteriorly (down in front) as this muscle contracts. The abdominal muscles must prevent this tendency by posteriorly rotating the
pelvis (pulling up in front), thus flattening the
lower back.

The sartorius, a two-joint muscle, is effective as
a hip flexor or as a knee flexor. It is weak when
both actions take place at the same time. Observe
that, in attempting to cross the knees when in a
sitting position, one customarily leans well back,
thus raising the origin to lengthen this muscle,
making it more effective in flexing and crossing
the knees. With the knees held extended, the sartorius becomes a more effective hip flexor. It is
the longest muscle in the body and is strengthened when hip flexion activities are performed as
described for developing the iliopsoas. Stretching
may be accomplished by a partner passively taking the hip into extreme extension, adduction,
and internal rotation with the knee extended.

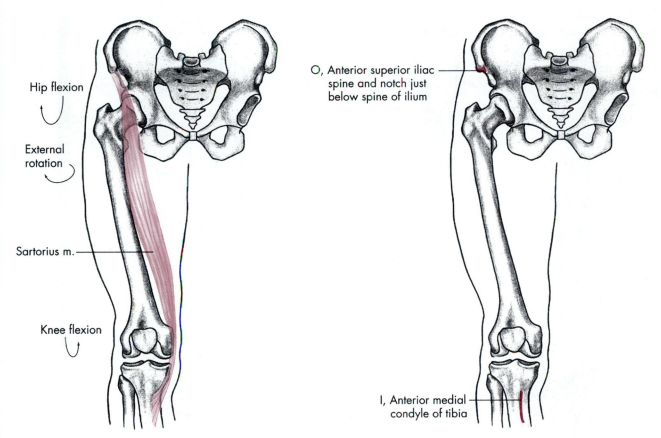

FIG. 7.8 ● Sartorius muscle, anterior view. *O,* Origin; *I,* insertion.

Rectus femoris muscle FIG. 7.9

(rek´tus fem´or-is)

Origin

Anterior inferior iliac spine of the ilium and groove (posterior) above the acetabulum

Insertion

Superior aspect of the patella and patellar tendon to the tibial tuberosity

Action

Flexion of the hip
Extension of the knee

Palpation

Any place on the anterior surface of the femur

Innervation

Femoral nerve (L2–4)

Application, strengthening, and flexibility

Pulling from the anterior inferior iliac spine of the ilium, the rectus femoris muscle has the same tendency to anteriorly rotate the pelvis (down in front and up in back). Only the abdominal muscles can prevent this from occurring. In speaking of the hip flexor group in general, it may be said that many people permit the pelvis to be permanently tilted forward as they get older. The relaxed abdominal wall does not hold the pelvis up; therefore, an increased lumbar curve results.

Generally, a muscle's ability to exert force decreases as it shortens. This explains why the rectus femoris muscle is a powerful extensor of the knee when the hip is extended but is weak when the hip is flexed. This muscle is exercised, along with the vastus group, in running, jumping, hopping, and skipping. In these movements, the hips are extended powerfully by the gluteus maximus and the hamstring muscles, which counteract the tendency of the rectus femoris muscle to flex the hip while it extends the knee. It can be remembered as one of the quadriceps muscle group. The rectus femoris is developed by performing hip flexion exercises or knee extension exercises against manual resistance.

The rectus femoris is stretched by fully flexing the knee while extending the hip.

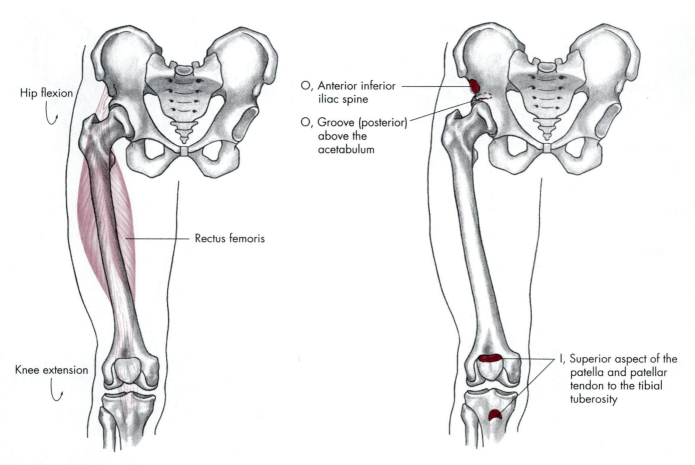

Hip flexion

Rectus femoris

Knee extension

O, Anterior inferior iliac spine

O, Groove (posterior) above the acetabulum

I, Superior aspect of the patella and patellar tendon to the tibial tuberosity

FIG. 7.9 ● Rectus femoris muscle, anterior view. *O,* Origin; *I,* insertion.

Tensor fasciae latae muscle   FIG. 7.10

(ten´sor fas´i-e la´te)

Origin

Anterior iliac crest and surface of the ilium just
below the crest

Insertion

One-fourth of the way down the thigh into the
iliotibial tract, which in turn inserts onto Gerdy's
tubercle of the anterolateral tibial condyle

Action

Abduction of the hip
Flexion of the hip
Tendency to rotate the hip internally as it flexes

Palpation

Slightly in front of the greater trochanter

Innervation

Superior gluteal nerve (L4–5, S1)

Application, strengthening, and flexibility

The tensor fasciae latae muscle aids in preventing
external rotation of the hip as it is flexed by other
flexor muscles.

The tensor fasciae latae muscle is used when
flexion and internal rotation take place. This is a
weak movement but is important in helping direct
the leg forward so that the foot is placed straight
forward in walking and running. Thus, from the
supine position, raising the leg with definite internal rotation of the femur will call it into action.

The tensor fasciae latae may be developed by
performing hip abduction exercises against gravity and resistance while in a side-lying position.
This is done simply by abducting the hip that is
up and then slowly lowering it back to rest against
the other leg. Stretch may be applied by remaining on the side and having a partner passively
move the downside hip into full extension,
adduction, and external rotation.

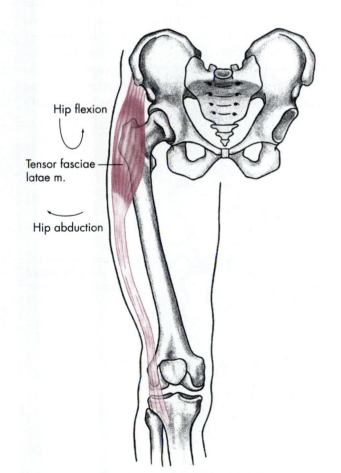

Hip flexion

Tensor fasciae latae m.

Hip abduction

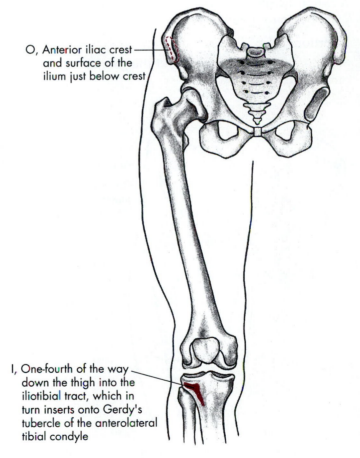

O, Anterior iliac crest
and surface of the
ilium just below crest

I, One-fourth of the way
down the thigh into the
iliotibial tract, which in
turn inserts onto Gerdy's
tubercle of the anterolateral
tibial condyle

FIG. 7.10 ● Tensor fasciae latae muscle, anterior view. *O*, Origin; *I*, insertion.

The six deep lateral rotator muscles—

piriformis
(pi-ri-for´mis),

gemellus superior
(je-mel´us su-pe´ri-or),

gemellus inferior
(je-mel´us in-fe´ri-or),

obturator externus
(ob-tu-ra´tor eks-ter´nus),

obturator internus
(ob-tu-ra´tor in-ter´nus),

quadratus femoris
(kwad-ra´tus fem´or-is)   FIG 7.11

Origin
Anterior sacrum, posterior portions of the ischium, and obturator foramen

Insertion
Superior and posterior aspect of the greater trochanter

Action
External rotation of the hip

Palpation
Cannot be palpated

Innervation
Piriformis: first or second sacral nerve (S1–2)
Gemellus superior: sacral nerve (L5, S1–2)
Gemellus inferior: branches from sacral plexus (L4–5, S1–2)
Obturator externus: obturator nerve (L3–4)
Obturator internus: branches from sacral plexus (L4–5, S1–2)
Quadratus femoris: branches from sacral plexus (L4–5, S1)

Application, strengthening, and flexibility

The six lateral rotators are used powerfully in movements of external rotation of the femur, as in sports in which the individual takes off on one leg from a preliminary internal rotation. Throwing a baseball and swinging a baseball bat, in which there is rotation of the hip, are typical examples.

Standing on one leg and forcefully turning the body away from that leg is accomplished by contraction of these muscles, and it may be repeated for strengthening purposes. A partner may provide resistance as development progresses. The six deep lateral rotators may be stretched in the supine position with a partner passively internally rotating and slightly flexing the hip.

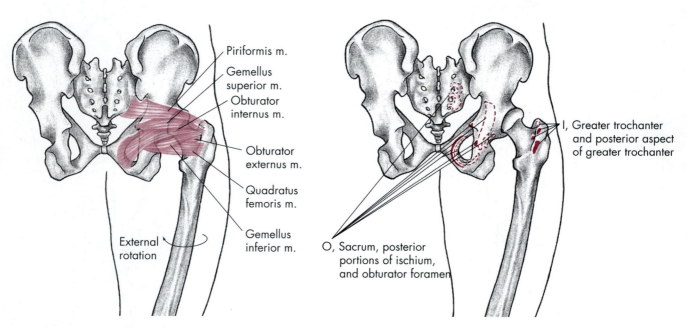

FIG. 7.11 ● The six deep lateral rotator muscles, posterior view: piriformis, gemellus superior, gemellus inferior, obturator externus, obturator internus, and quadratus femoris.

Gluteus minimus muscle FIG. 7.12

(glu´te-us min´i-mus)

Origin

Lateral surface of the ilium just below the origin of
the gluteus medius

Insertion

Anterior surface of the greater trochanter of the
femur

Action

Abduction of the hip
Internal rotation as the femur abducts

Palpation

Cannot be palpated

Innervation

Superior gluteal nerve (L4–5, S1)

Application, strengthening, and flexibility

Both the gluteus minimus and the gluteus medius
are used in powerfully maintaining proper hip
abduction while running. As a result, both of
these muscles are exercised effectively in running,
hopping, and skipping, in which weight is transferred forcefully from one foot to the other. As the
body ages, the gluteus medius and gluteus minimus muscles tend to lose their effectiveness. The
spring of youth, as far as the hips are concerned,
resides in these muscles. To have great drive in
the legs, these muscles must be fully developed.

The gluteus minimus is best strengthened by
performing hip abduction exercises similar to the
ones described for the tensor fasciae latae and
gluteus medius muscles. It may also be developed
by performing hip internal rotation exercises
against manual resistance. Stretching of this muscle is accomplished by extreme hip adduction
with slight external rotation.

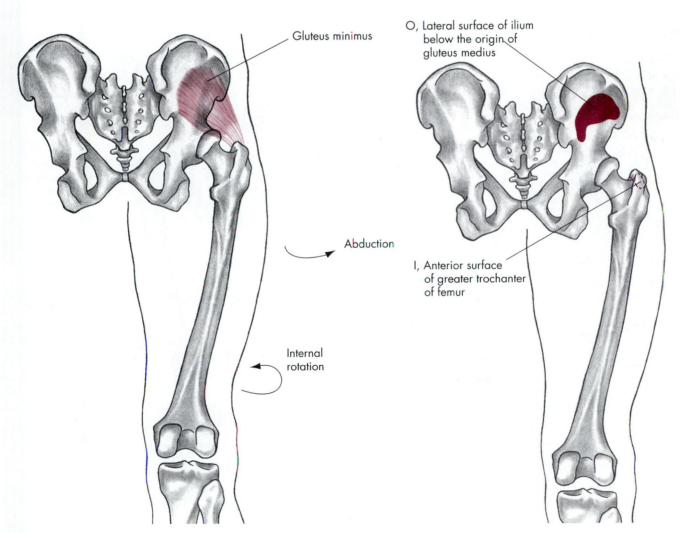

FIG. 7.12 • Gluteus minimus muscle, posterior view. *O*, Origin; *I*, insertion.

127

Gluteus medius muscle FIG. 7.13

(glu´te-us me´di-us)

Origin

Lateral surface of the ilium just below the crest

Insertion

Posterior and middle surfaces of the greater
  trochanter of the femur

Action

Abduction of the hip
External rotation as the hip abducts (posterior
  fibers)
Internal rotation (anterior fibers)

Palpation

Slightly in front of and a few inches above the
  greater trochanter

Innervation

Superior gluteal nerve (L4–5, S1)

Application, strengthening, and flexibility

Typical action of the gluteus medius and gluteus
minimus muscles is seen in walking. As the
weight of the body is suspended on one leg, these
muscles prevent the opposite pelvis from sagging.
Weakness in the gluteus medius and gluteus
minimus can result in the Trendelenburg gait.
With this weakness, the individual's opposite
pelvis will sag on weight bearing because the hip
abductors cannot maintain proper alignment.

Hip external rotation exercises performed
against resistance can provide some strengthening
for the gluteus medius, but it is best strengthened
by performing the side-lying leg raises or hip
abduction exercises as described for the tensor
fasciae latae. The gluteus medius is best stretched
by moving the hip into extreme adduction in front
of the opposite extremity and then behind it.

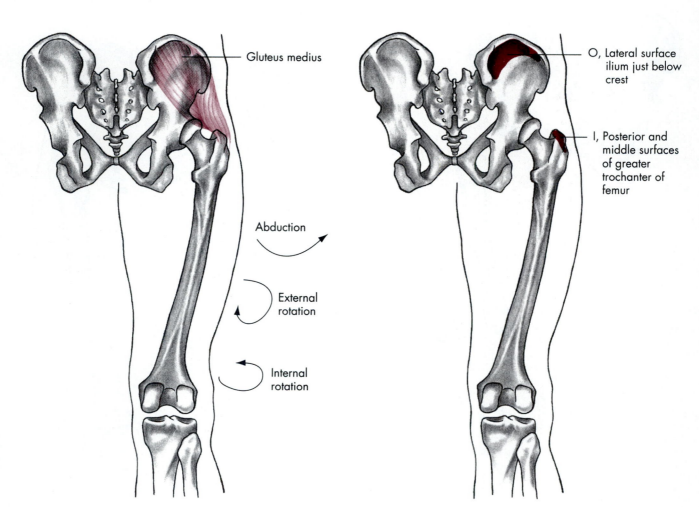

FIG. 7.13 • Gluteus medius muscle, posterior view. *O*, Origin; *I*, insertion.

Gluteus maximus muscle   FIG. 7.14

(glu´te-us maks´i-mus)

Origin

Posterior one-fourth of the crest of the ilium, posterior surface of the sacrum and coccyx near the ilium, and fascia of the lumbar area

Insertion

Oblique ridge on the lateral surface of the greater trochanter and the iliotibial band of the fasciae latae

Action

Extension of the hip
External rotation of the hip
Lower fibers, which assist in adduction

Palpation

Wide area on the posterior surface of the pelvis

Innervation

Inferior gluteal nerve (L5, S1–2)

Application, strengthening, and flexibility

The gluteus maximus muscle comes into action when movement between the pelvis and the femur approaches and goes beyond 15 degrees of extension. As a result, it is not used extensively in ordinary walking. It is important in extension of the thigh with external rotation.

Strong action of the gluteus maximus muscle is seen in running, hopping, skipping, and jumping. Powerful extension of the thigh is secured in the return to standing from a squatting position, especially if a barbell with weights is placed on the shoulders.

Hip extension exercises from a forward-leaning or prone position may be used to develop this muscle. This muscle is most emphasized when the hip starts from a flexed position and moves to full extension with the knee flexed 30 degrees or more to reduce the hamstrings' involvement in the action.

The gluteus maximus is stretched in the supine position with full hip flexion to the ipsilateral axilla and then to the contralateral axilla with the knee in flexion. Simultaneous internal hip rotation accentuates this stretch.

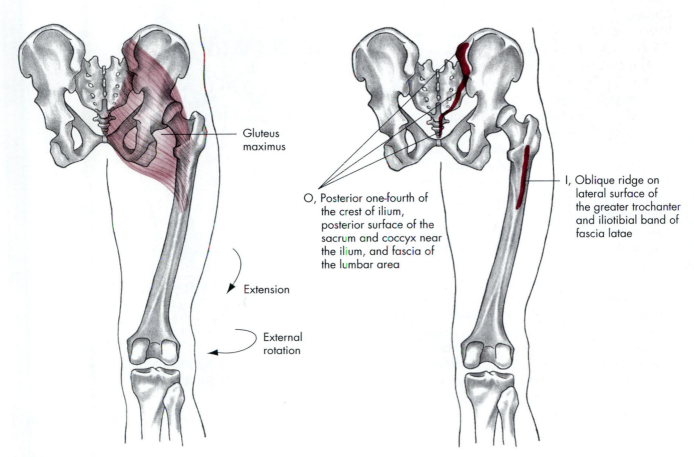

Gluteus maximus

Extension

External rotation

O, Posterior one-fourth of the crest of ilium, posterior surface of the sacrum and coccyx near the ilium, and fascia of the lumbar area

I, Oblique ridge on lateral surface of the greater trochanter and iliotibial band of fascia latae

FIG. 7.14 ● Gluteus maximus muscle, posterior view. *O,* Origin; *I,* insertion.

Semitendinosus muscle   FIG. 7.15

(sem´i-ten-di-no´sus)

Origin

Ischial tuberosity

Insertion

Upper anterior medial surface of the tibia

Action

Extension of the hip
Flexion of the knee
Internal rotation of the hip
Internal rotation of the knee

Palpation

Near the knee on the posteromedial side

Innervation

Sciatic nerve—tibial division (L5, S1–2)

Application, strengthening, and flexibility

This two-joint muscle is most effective when contracting to either extend the hip or flex the knee. When there is extension of the hip and flexion of the knee at the same time, both movements are weak. When the trunk is bent forward with the knees straight, the hamstring muscles have a powerful pull on the rear pelvis and tilt it down in back by full contraction. If the knees are flexed when this movement takes place, one can observe that the work is done chiefly by the gluteus maximus muscle.

On the other hand, when the muscles are used in powerful flexion of the knees, as in hanging by the knees from a bar, the flexors of the hip come into play to raise the origin of these muscles and make them more effective as knee flexors. By full extension of the hips in this movement, the knee flexion movement is weakened. These muscles are used in ordinary walking as extensors of the hip and allow the gluteus maximus to relax in the movement.

The semitendinosus is best developed through hamstring curls as described for the biceps femoris, but it is emphasized more if the knee is maintained in internal rotation throughout the range of motion, which brings the origin and insertion more in line with each other. The semitendinosus is stretched by maximally extending the knee while flexing the externally rotated and slightly abducted hip.

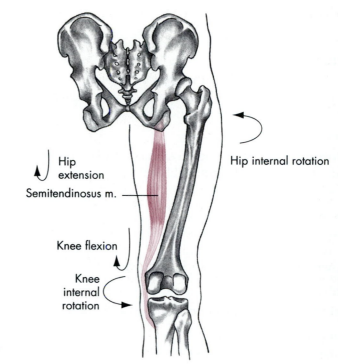

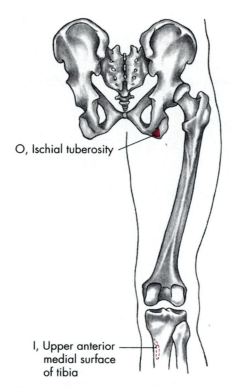

FIG. 7.15 • Semitendinosus muscle, posterior view. *O*, Origin; *I*, insertion.

Semimembranosus muscle   FIG. 7.16

(sem´i-mem´bra-no´sus)

Origin

Ischial tuberosity

Insertion

Posteromedial surface of the medial tibial condyle

Action

Extension of the hip
Flexion of the knee
Internal rotation of the hip
Internal rotation of the knee

Palpation

Largely covered by other muscles, the tendon can
be felt at the posterior aspect of the tibia on the
medial side

Innervation

Sciatic nerve—tibial division (L5, S1–2)

Application, strengthening, and flexibility

Both the semitendinosus and semimembranosus
are responsible for internal rotation of the knee,
along with the popliteus muscle, which is discussed in the next chapter. Because of the manner
in which they cross the joint, the muscles are very
important in providing dynamic medial stability to
the knee joint.

The semimembranosus is best developed by
performing leg curls. Internal rotation of the knee
throughout the range accentuates the activity of
this muscle. The semimembranosus is stretched in
the same manner as the semitendinosus.

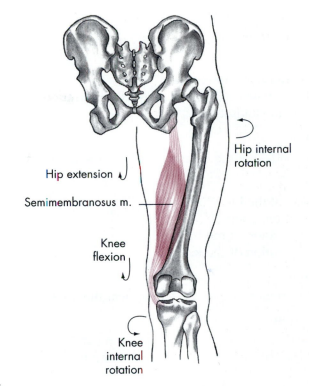

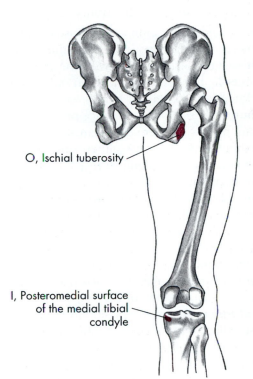

FIG. 7.16 ● Semimembranosus muscle, posterior
view. *O,* Origin; *I,* insertion.

Biceps femoris muscle   FIG. 7.17

(bi´seps fem´or-is)

Origin

Long head: ischial tuberosity
Short head: lower half of the linea aspera, and
   lateral condyloid ridge

Insertion

Lateral condyle of the tibia and head of the fibula

Action

Extension of the hip
Flexion of the knee
External rotation of the hip
External rotation of the knee

Palpation

Lateral posterior side of the femur, near the knee

Innervation

Long head: sciatic nerve—tibial division (S1–3)
Short head: sciatic nerve—peroneal division
   (L5, S1–2)

Application, strengthening, and flexibility

The semitendinosus, semimembranosus, and biceps femoris muscles are known as the hamstrings. These muscles, together with the gluteus maximus muscle, are used in extension of the hip when the knees are straight or nearly so. Thus, in running, jumping, skipping, and hopping, these muscles are used together. The hamstrings are used without the aid of the gluteus maximus, however, when one is hanging from a bar by the knees. Similarly, the gluteus maximus is used without the aid of the hamstrings when the knees are flexed while the hips are being extended. This occurs when rising from a knee-bend position to a standing position.

The biceps femoris is best developed through knee flexion exercises against resistance. Commonly known as hamstring curls or leg curls, they may be performed in a prone position on a knee table or standing with ankle weights attached. This muscle is emphasized when performing hamstring curls while attempting to maintain the knee joint in external rotation. This externally rotated position brings its insertion in alignment with its origin.

The biceps femoris is best stretched by maximally extending the knee while flexing the internally rotated and slightly adducted hip.

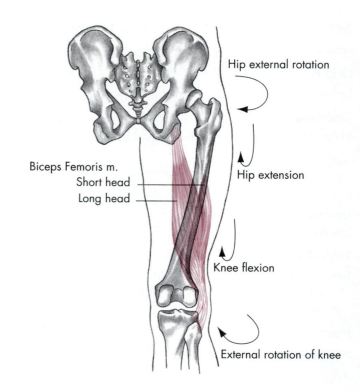

Biceps Femoris m.
   Short head
   Long head

Hip external rotation

Hip extension

Knee flexion

External rotation of knee

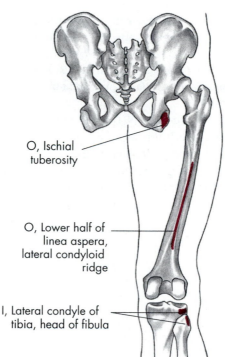

O, Ischial tuberosity

O, Lower half of linea aspera, lateral condyloid ridge

I, Lateral condyle of tibia, head of fibula

FIG. 7.17 ● Biceps femoris muscle, posterior view. *O*, Origin; *I*, insertion.

Adductor brevis muscle   FIG. 7.18

(ad-duk´tor bre´vis)

Origin

Front of the inferior pubic ramus just below the
  origin of the longus

Insertion

Lower two-thirds of the pectineal line of the femur
  and the upper half of the medial lip of the linea
  aspera

Action

Adduction of the hip
External rotation as it adducts the hip

Palpation

Cannot be palpated

Innervation

Obturator nerve (L3–4)

Application, strengthening, and flexibility

The adductor brevis muscle, along with the other
adductor muscles, provides powerful movement
of the thighs toward each other. Squeezing the
legs together toward each other against resistance
is effective in strengthening the adductor brevis.
Abducting the extended and internally rotated hip
provides stretching of the adductor brevis.

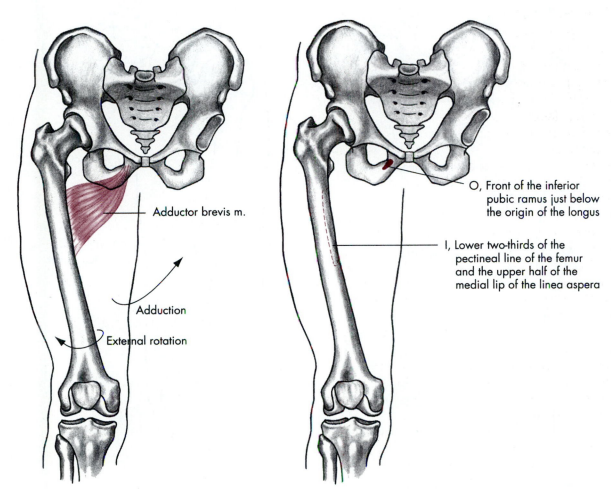

Adductor brevis m.

Adduction

External rotation

O, Front of the inferior
pubic ramus just below
the origin of the longus

I, Lower two-thirds of the
pectineal line of the femur
and the upper half of the
medial lip of the linea aspera

FIG. 7.18 ● Adductor brevis muscle, anterior view. *O,* Origin; *I,* insertion.

Adductor longus muscle   FIG. 7.19

(ad-duk'tor long'gus)

Origin

Anterior pubis just below its crest

Insertion

Middle third of the linea aspera

Action

Adduction of the hip

Assists in flexion of the hip

Palpation

Just below the pubic bone on the medial side

Innervation

Obturator nerve (L3–4)

Application, strengthening, and flexibility

The muscle may be strengthened by using the scissors exercise, which requires the subject to sit on the floor with the legs spread wide while a partner puts his or her legs or arms inside each lower leg to provide resistance. As the subject attempts to adduct his or her legs together, the partner provides manual resistance throughout the range of motion. This exercise may be used for either one or both legs. The adductor longus is stretched in the same manner as the adductor brevis.

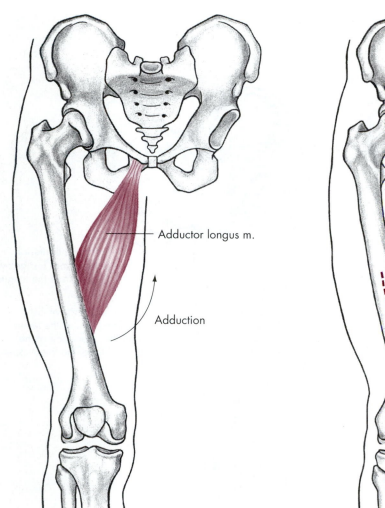

Adductor longus m.

Adduction

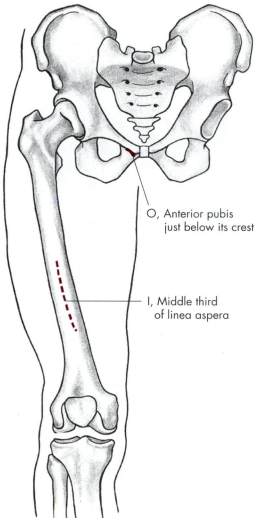

O, Anterior pubis just below its crest

I, Middle third of linea aspera

FIG. 7.19 ● Adductor longus muscle, anterior view. *O*, Origin; *I*, insertion.

Modified from Anthony CP, Kolthoff NJ: *Textbook of anatomy and physiology*, ed 9, St. Louis, 1975, Mosby.

Adductor magnus muscle   FIG. 7.20

(ad-duk´tor mag´nus)

Origin

Edge of the entire ramus of the pubis and the ischium and ischial tuberosity

Insertion

Whole length of the linea aspera, inner condyloid ridge, and adductor tubercle

Action

Adduction of the hip

External rotation as the hip adducts

Palpation

Posteromedial surface of the thigh

Innervation

Anterior: obturator nerve (L2–4)

Posterior: sciatic nerve (L4–5, S1–3)

Application, strengthening, and flexibility

The adductor magnus muscle is used in the breaststroke kick in swimming or in horseback riding. Since the adductor muscles (adductor magnus, adductor longus, adductor brevis, and gracilis muscles) are not heavily used in ordinary movement, some prescribed activity for them should be provided. Some modern exercise equipment is engineered to provide resistance for hip adduction movement. Hip adduction exercises such as those described for the adductor brevis and the adductor longus may be used for strengthening the adductor magnus as well. The adductor magnus is stretched in the same manner as the adductor brevis and adductor longus.

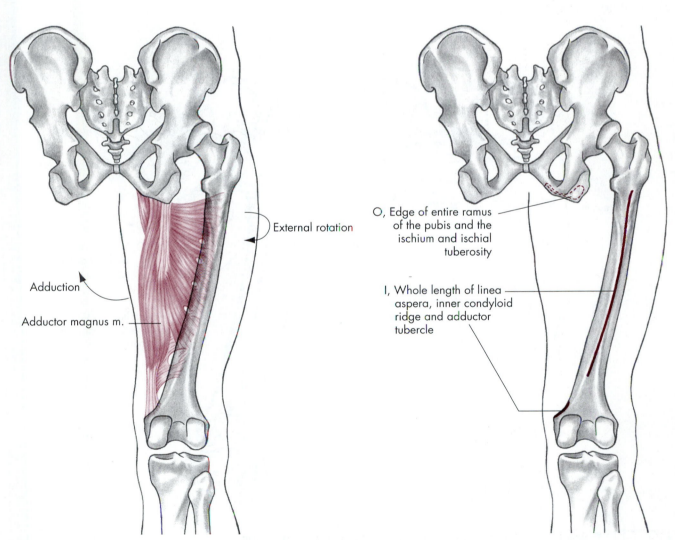

External rotation

Adduction

Adductor magnus m.

O, Edge of entire ramus of the pubis and the ischium and ischial tuberosity

I, Whole length of linea aspera, inner condyloid ridge and adductor tubercle

FIG. 7.20 ● Adductor magnus muscle, posterior view. *O*, Origin; *I*, insertion.

Pectineus muscle FIG. 7.21

(pek-tin´e-us)

Origin

Space 1 inch wide on the front of the pubis just above the crest

Insertion

Rough line leading from the lesser trochanter down to the linea aspera

Action

Flexion of the hip
Adduction of the hip
Internal rotation of the hip

Palpation

Angle between the pubic bone and the femur; hard to distinguish from the adductor longus muscle

Innervation

Femoral nerve (L2–4)

Application, strengthening, and flexibility

As the pectineus contracts, it also tends to rotate the pelvis anteriorly. The abdominal muscles pulling up on the pelvis in front prevent this tilting action.

The pectineus muscle is exercised together with the iliopsoas muscle in leg raising and lowering. Hip flexion exercises and hip adduction exercises against resistance may be used for strengthening this muscle.

The pectineus is stretched by fully abducting the extended and externally rotated hip.

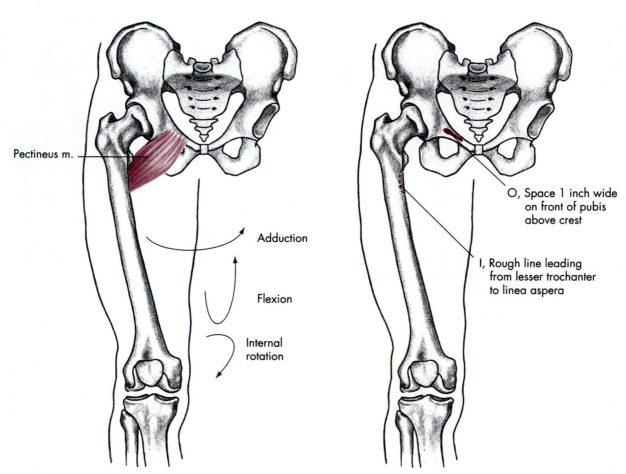

Pectineus m.

Adduction

Flexion

Internal rotation

O, Space 1 inch wide on front of pubis above crest

I, Rough line leading from lesser trochanter to linea aspera

FIG. 7.21 ● Pectineus muscle, anterior view. *O*, Origin; *I*, insertion.

Modified from Anthony CP, Kolthoff NJ: *Textbook of anatomy and physiology*, ed 9, St. Louis, 1975, Mosby.

Gracilis muscle   FIG. 7.22

(gras´-il-is)

Origin

Anteromedial edge of the descending ramus of the
  pubis

Insertion

Anterior medial surface of the tibia below the
  condyle

Action

Adduction of the hip
Flexion of the knee
Internal rotation of the hip

Palpation

Medial side of the thigh 2 to 3 inches below the
  pubic bone

Innervation

Obturator nerve (L2–4)

Application, strengthening, and flexibility

The gracilis muscle performs the same function as
the other adductors but adds some weak assistance to knee flexion.

The adductor muscles as a group (adductor
magnus, adductor longus, adductor brevis, and
gracilis) are called into action in horseback riding
and in doing the breaststroke kick in swimming.
Proper development of the adductor group prevents soreness after participation in these sports.
The gracilis is strengthened with the same exercises as described for the other hip adductors. The
gracilis may be stretched in a manner similar to
the adductors, except that the knee must be
extended.

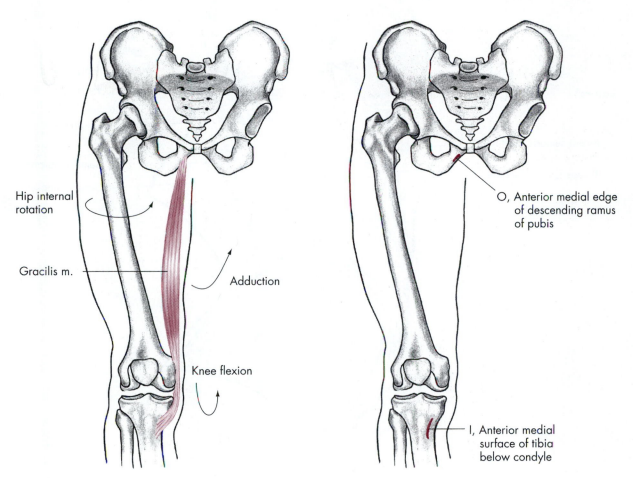

Hip internal rotation

Gracilis m.

Adduction

Knee flexion

O, Anterior medial edge
of descending ramus
of pubis

I, Anterior medial
surface of tibia
below condyle

FIG. 7.22 ● Gracilis muscle, anterior view. *O,* Origin; *I,* insertion.

Modified from Anthony CP, Kolthoff NJ: *Textbook of anatomy and physiology,* ed 9, St. Louis, 1975, Mosby.

Muscle identification

In developing a thorough and practical knowledge of the muscular system, it is essential that individual muscles be understood. Figs. 7.23 and 7.24 illustrate groups of muscles that work together to produce joint movement.

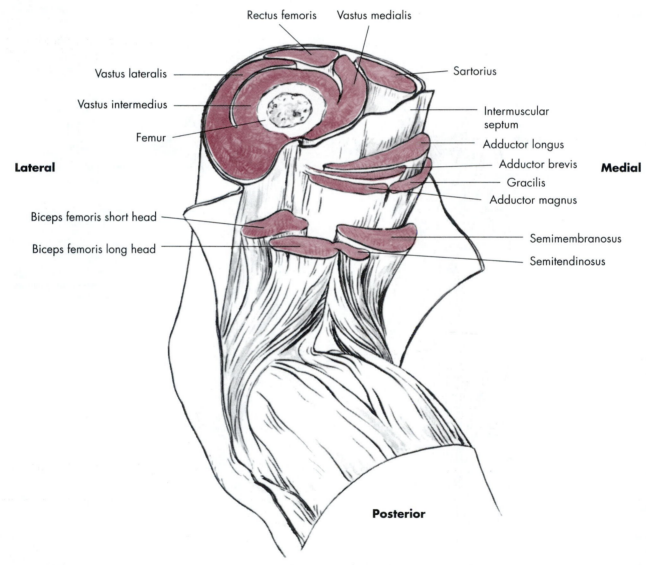

FIG. 7.23 • Cross section of the left thigh at the midsection.

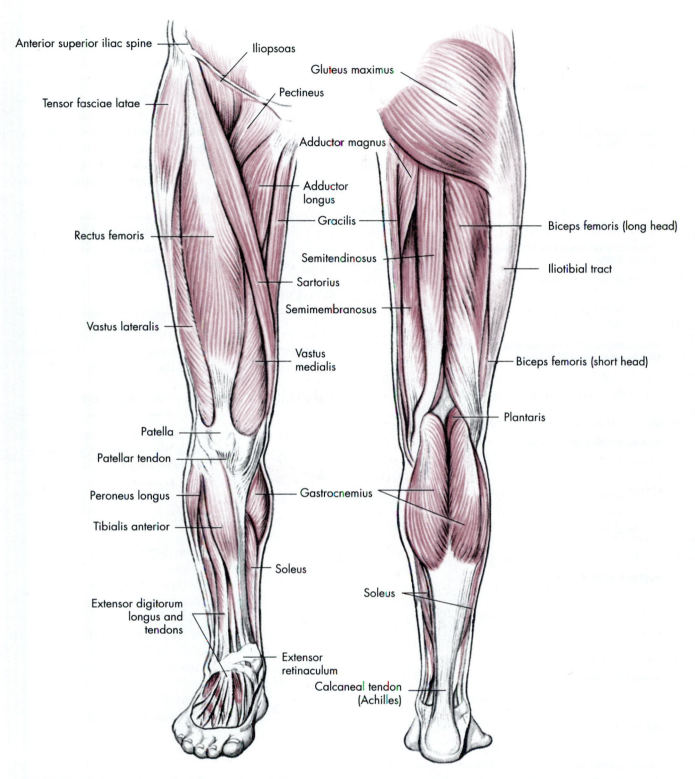

FIG. 7.24 ● **Left**, Superficial muscles of the right upper leg, anterior surface; **right**, superficial muscles of the right upper leg, posterior surface.

Modified from Anthony CP, Kolthoff NJ: *Textbook of anatomy and physiology*, ed 9, St. Louis, 1975, Mosby.

Web sites

Anatomy & Physiology Tutorials:

www.gwc.maricopa.edu/class/bio201/index.htm

Radiologic Anatomy Browser:

radlinux1.usuf1.usuhs.mil/rad/iong/ index.html

This site has numerous radiological views of the musculoskeletal system.

University of Arkansas Medical School Gross Anatomy for Medical Students:

anatomy.uams.edu/htmlpages/anatomyhtml/ gross.html

Dissections, anatomy tables, atlas images, links, etc.

Loyola University Medical Center: Structure of the Human Body:

www.meddean.luc.edu/lumen/MedEd/ GrossAnatomy/GA.html

An excellent site with many slides, dissections, tutorials, etc., for the study of human anatomy.

Wheeless' Textbook of Orthopaedics:

www.medmedia.com/

This site has an extensive index of links to the fractures, joints, muscles, nerves, trauma, medications, medical topics, and lab tests, as well as links to orthopedic journals and other orthopedic and medical news.

Premiere Medical Search Engine:

www.medsite.com

This site allows the reader to enter any medical condition and it will search the net to find relevant articles.

Virtual Hospital:

www.vh.org

Numerous slides, patient information, etc.

Dynamic Human version 2.0 CD-ROM: The Visual Guide to Anatomy & Physiology:

www.mhhe.com/biosci/ap/dynamichuman2/

Web site that accompanies this CD-ROM.

Dynamic Human CD activities

1. Review anatomical landmarks as well as origins/insertions for the hip joint and pelvic girdle by clicking on **skeletal, anatomy**, and **gross anatomy**, then **pelvic girdle**. Also click on **skeletal, anatomy, gross anatomy, lower limbs, femur**, and **tibia/fibula**.
2. Review each of the muscles from this chapter by clicking on **muscular, anatomy, body regions**, and **thigh**.
3. What muscular rehabilitation do you think will be necessary following bony healing of a fractured femur? Click on **skeletal, clinical concepts**, and **fractured femur**.

Worksheet exercises

As an aid to learning, for in-class or out-of-class assignments, or for testing, tear-out worksheets are found at the end of the text (pp. 261 and 262).

Laboratory and review exercises

1. Locate the following parts of the pelvic girdle and hip joint on a human skeleton and on a subject:
   a. Skeleton
      1. Ilium
      2. Ischium
      3. Pubis
      4. Symphysis pubis
      5. Acetabulum
      6. Rami (ascending and descending)
      7. Obturator foramen
      8. Ischial tuberosity
      9. Anterior superior iliac spine
      10. Greater trochanter
      11. Lesser trochanter
   b. Subject
      1. Crest of ilium
      2. Anterior superior iliac spine
      3. Ischial tuberosity
      4. Greater trochanter

2. How and where can the following muscles be palpated on a human subject?
   a. Gracilis
   b. Sartorius
   c. Gluteus maximus
   d. Gluteus medius
   e. Gluteus minimus
   f. Biceps femoris
   g. Rectus femoris
   h. Semimembranosus
   i. Semitendinosus
   j. Adductor magnus
   k. Adductor longus
   l. Adductor brevis
3. Be prepared to indicate on a human skeleton, using a long rubber band, where each muscle has its origin and insertion.
4. Distinguish between hip flexion and trunk flexion.
5. Demonstrate the movement and list the muscles primarily responsible for the following hip movements:
   a. Flexion
   b. Extension
   c. Adduction
   d. Abduction
   e. External rotation
   f. Internal rotation
6. List the planes in which each of the following hip joint movements occur. List the respective axis of rotation for each movement in each plane.
   a. Flexion
   b. Extension
   c. Adduction
   d. Abduction
   e. External rotation
   f. Internal rotation
7. How is walking different from running in relation to the use of the hip joint muscle actions and the range of motion?
8. How may the walking gait be affected by a weakness in the gluteus medius muscle? Have a laboratory partner demonstrate the gait pattern associated with gluteus medius weakness. What is the name of this dysfunctional gait?
9. How might bilateral iliopsoas tightness affect the posture and movement of the lumbar spine in the standing position? Demonstrate and discuss this effect with a laboratory partner.
10. How might bilateral hamstring tightness affect the posture and movement of the lumbar spine in the standing position? Demonstrate and discuss this effect with a laboratory partner.
11. The hip joint and pelvic girdle muscles are listed at the left of the muscle analysis chart. Place a check in the column for each action of the muscle. Add a "P" for primary action.
12. Fill in the antagonistic muscle action chart by listing the muscle(s) or parts of muscles that are antagonistic in their actions to the muscles in the left column.

Muscle analysis chart • Hip joint and pelvic girdle

Muscles	Flexion	Extension	Abduction	Adduction	External rotation	Internal rotation
Gluteus maximus						
Gluteus medius						
Gluteus minimus						
Biceps femoris						
Semimem- branosus						
Semitendinosus						
Adductor magnus						
Adductor longus						
Adductor brevis						
Gracilis						
Lateral rotators						
Rectus femoris						
Sartorius						
Pectineus						
Iliopsoas						
Tensor fasciae latae						

Antagonistic muscle action chart • Hip joint and pelvic girdle

Agonist	Antagonist
Gluteus maximus	
Gluteus medius	
Gluteus minimas	
Biceps femoris	
Semimembranosus/ semitendinosus	
Adductor magnus/ adductor brevis	
Adductor longus	
Gracilis	
Lateral rotators	
Rectus femoris	
Sartorius	
Pectineus	
Iliopsoas	
Tensor fasciae latae	

References

Hislop HJ, Montgomery J: *Daniels and Worthingham's muscle testing: techniques of manual examination,* ed 6, Philadelphia, 1995, Saunders.

Kendall FP, McCreary EK, Provance, PG: *Muscles: testing and function,* ed 4, Baltimore, 1993, Lippincott Williams & Wilkins.

Lindsay DT: *Functional human anatomy,* St. Louis, 1996, Mosby.

Luttgens K, Hamilton N: *Kinesiology: scientific basis of human motion,* ed 9, Madison, WI, 1997, Brown & Benchmark.

Lysholm J, Wikland J: Injuries in runners, *American Journal of Sports Medicine* 15:168, September-October 1986.

Noahes TD, et al: Pelvic stress fractures in long distance runners, *American Journal of Sports Medicine* 13:120, March-April 1985.

Perreira J: Treating the quadriceps contusion, *Scholastic Coach* 57:38, October 1987.

Seeley RR, Stephens TD, Tate P: *Anatomy & physiology,* ed 2, St. Louis, 1992, Mosby-Year Book.

Sieg KW, Adams SP: *Illustrated essentials of musculoskeletal anatomy,* ed 2, Gainesville, FL, 1985, Megabooks.

Stone RJ, Stone JA: *Atlas of the skeletal muscles,* 1990, McGraw-Hill Companies Inc., New York.

Thibodeau GA, Patton KT: *Anatomy & physiology,* ed 9, St. Louis, 1993, Mosby.

The knee joint

8

Objectives

- To identify on a human skeleton selected bony features of the knee

- To explain the cartilaginous and ligamentous structures of the knee joint

- To draw and label on a skeletal chart muscles and ligaments of the knee joint

- To palpate the superficial knee joint structures and muscles on a human subject

- To demonstrate and palpate with a fellow student all the movements of the knee joint and list their respective planes of motion and axes of rotation

- To name and explain the actions and importance of the quadriceps and hamstring muscles

- To list and organize the muscles that produce the movements of the knee joint and list their antagonists

The knee joint is the largest joint in the body and is very complex. It is primarily a hinge joint. The combined functions of weight bearing and locomotion place considerable stress and strain on the knee joint. Powerful knee joint extensor and flexor muscles, combined with a strong ligamentous structure, provide a strong functioning joint in most instances.

Bones

The enlarged femoral condyles articulate on the enlarged condyles of the tibia, somewhat in a horizontal line. Since the femur projects downward at an oblique angle toward the midline, its medial condyle is slightly larger than the lateral condyle.

The top of the medial and lateral tibial condyles, known as the medial and lateral tibial plateaus, serve as receptacles for the femoral condyles. The tibia is the medial bone in the leg and bears much more of the body's weight than the fibula. The fibula serves as the attachment for some very important knee joint structures, although it does not articulate with the femur or patella and is not part of the knee joint.

The patella is a sesamoid (floating) bone contained within the quadriceps muscle group and patellar tendon. Its location allows it to serve the quadriceps in a fashion similar to a pulley by creating an improved angle of pull. This results in a greater mechanical advantage when performing knee extension.

Joints   FIG. 8.1

The knee joint proper, or tibiofemoral joint, is classified as a ginglymus joint because it functions like a hinge. It moves between flexion and extension without side-to-side movement into abduction or adduction. However, it is sometimes referred to as a trochoginglymus joint because of the internal and external rotation movements that can occur during flexion. Some authorities argue that it should be classified as a condyloid joint due to its structure. The patellofemoral joint is classified as an arthrodial joint due to the gliding nature of the patella on the femoral condyles.

The ligaments provide static stability to the knee joint, and contractions of the quadriceps and hamstrings produce dynamic stability. The surfaces between the femur and tibia are protected by articular cartilage, as is true of all diarthrodial joints. In addition to the articular cartilage covering the ends of the bones, there are specialized cartilages (see Fig. 8.1), known as the menisci, that form cushions between the bones. These menisci are attached to the tibia and deepen the tibial fossa, thereby enhancing stability.

The medial semilunar cartilage or, more technically, the medial meniscus, is located on the medial tibial plateau to form a receptacle for the medial femoral condyle. The lateral semilunar cartilage (lateral meniscus) sits on the lateral tibial plateau to receive the lateral femoral condyle. Both of these menisci are thicker on the outside border and taper down very thin to the inside border. They can slip about slightly and are held in place by various small ligaments. The medial meniscus is the larger of the two and has a much more open C appearance than the rather closed C lateral meniscus configuration. Either or both of the menisci may be torn in several different areas from a variety of mechanisms, resulting in varying degrees of severity and problems. These injuries often occur due to the significant compression and shear forces that develop as the knee rotates while flexing or extending during quick directional changes in running.

Two very important ligaments of the knee are the anterior and posterior cruciate, so named because they cross within the knee between the tibia and the femur. These ligaments are vital in respectively maintaining the anterior and posterior stability of the knee joint, as well as the rotatory stability (see Fig. 8.1).

The anterior cruciate ligament (ACL) tear is one of the most common serious injuries to the knee. The mechanism of this injury often involves noncontact rotary forces associated with planting and cutting. Studies have also shown that the ACL may be disrupted in a hyperextension mechanism or solely by a violent contraction of the quadriceps which pulls the tibia forward on the femur. Fortunately, the posterior cruciate ligament (PCL) is not often injured. Injuries of the posterior cruciate usually come about through direct contact with an opponent or with the playing surface.

On the medial side of the knee is the tibial (medial) collateral ligament (MCL) (see Fig. 8.1), which maintains medial stability by resisting valgus forces or preventing the knee joint from being abducted. Injuries to the tibial collateral occur quite commonly, particularly in contact or collision sports in which a teammate or an opponent falls against the lateral aspect of the knee or leg, causing medial opening of the knee joint and stress to the medial ligamentous structures.

On the lateral side of the knee, the fibular (lateral) collateral ligament (LCL) joins the fibula and the femur. Injuries to this ligament are infrequent.

In addition to the other intraarticular ligaments detailed in Fig. 8.1, there are numerous other ligaments not shown that are contiguous with the joint capsule. Generally, these ligaments are of lesser importance and will not be discussed further.*

The knee joint is well supplied with synovial fluid from the synovial cavity, which lies under the patella and between the surfaces of the tibia and the femur. Commonly, this synovial cavity is called the "capsule of the knee." More than 10 bursae are located in the knee, some of which are connected to the synovial cavity. Bursae are located where they can absorb shock or prevent friction.

The knee can usually extend to 180 degrees or a straight line, although it is not uncommon for some knees to hyperextend up to 10 degrees or more. When the knee is in full extension, it can move from there to about 140 degrees of flexion. With the knee flexed 30 degrees or more, approximately 30 degrees of internal rotation and 45 degrees of external rotation can occur.

---

*More detailed discussion of the knee is found in anatomy texts and athletic training manuals.

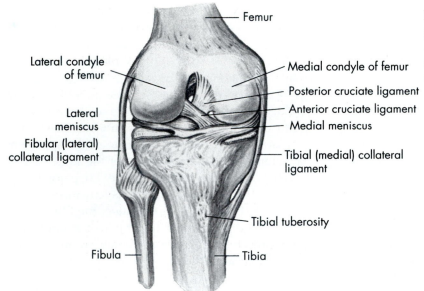

Femur

Lateral condyle of femur

Medial condyle of femur

Posterior cruciate ligament

Anterior cruciate ligament

Lateral meniscus

Medial meniscus

Fibular (lateral) collateral ligament

Tibial (medial) collateral ligament

Tibial tuberosity

Fibula

Tibia

Anterior view with patella removed

FIG. 8.1 ● Ligaments and menisci of the right knee.

Modified from Anthony CP, Kolthoff NJ: *Textbook of anatomy and physiology*, ed 9, St. Louis, 1975, Mosby.

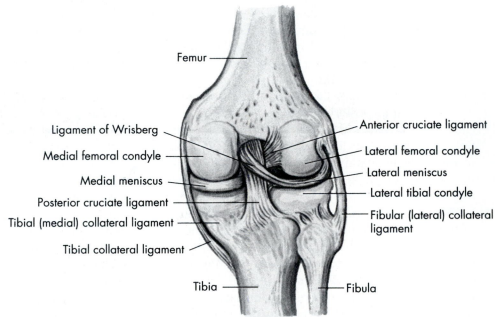

Femur

Ligament of Wrisberg

Anterior cruciate ligament

Medial femoral condyle

Lateral femoral condyle

Medial meniscus

Lateral meniscus

Posterior cruciate ligament

Lateral tibial condyle

Tibial (medial) collateral ligament

Fibular (lateral) collateral ligament

Tibial collateral ligament

Tibia

Fibula

Posterior view

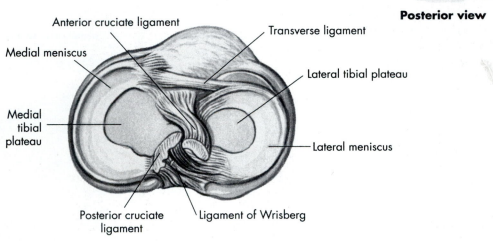

Anterior cruciate ligament

Transverse ligament

Medial meniscus

Lateral tibial plateau

Medial tibial plateau

Lateral meniscus

Posterior cruciate ligament

Ligament of Wrisberg

Superior view with femur removed

146

Movements   FIG. 8.2

Flexion and extension of the knee occur in the sagittal plane, whereas internal and external rotation occur in the horizontal plane. The knee will not allow rotation unless it is flexed 20 to 30 degrees or more.

Flexion: bending or decreasing the angle between the femur and lower leg characterized by the heel moving toward the buttocks

Extension: straightening or increasing the angle between the femur and the lower leg

External rotation: rotary movement of the lower leg laterally away from the midline

Internal rotation: rotary movement of the lower leg medially toward the midline

FIG. 8.2 ● Movements of the knee with prime movers illustrated.

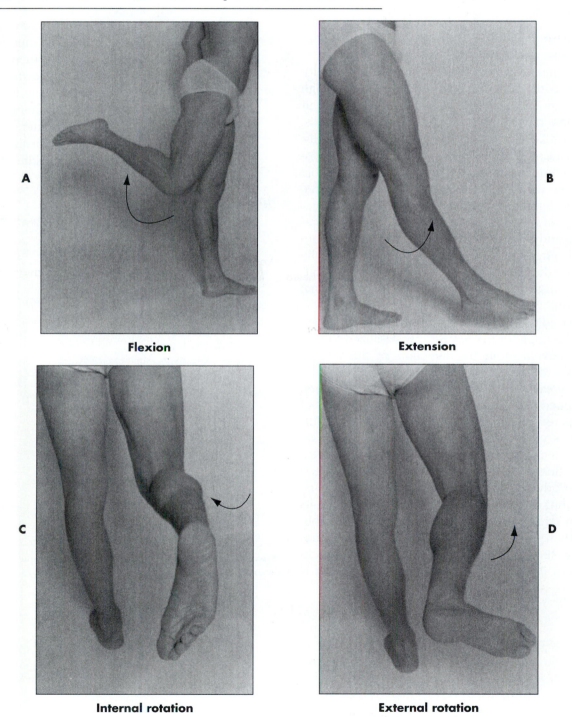

A   **Flexion**

B   **Extension**

C   **Internal rotation**

D   **External rotation**

Muscles

Some of the muscles involved in knee joint movements were discussed in Chapter 7 because of their biarticular arrangement with both the hip and knee joints. As a result, they will not be covered again fully in this chapter. The knee joint muscles that have already been addressed are

Knee extensor: rectus femoris
Knee flexors: sartorius, biceps femoris,
   semitendinosus, semimembranosus,
   and gracilis

The gastrocnemius muscle, discussed in Chapter 9, also assists minimally with knee flexion.

The muscle group that extends the knee is located in the anterior compartment of the thigh and is known as the quadriceps. It consists of four muscles: the rectus femoris, the vastus lateralis, the vastus intermedius, and the vastus medialis. The hamstring muscle group is located in the posterior compartment of the thigh and is responsible for knee flexion. The hamstrings consist of three muscles: the semitendinosus, the semimembranosus, and the biceps femoris. The semimembranosus and semitendinosus muscles (medial hamstrings) are assisted by the popliteus in internally rotating the knee, whereas the biceps femoris (lateral hamstring) is responsible for knee external rotation.

Two-joint muscles are most effective when either the origin or the insertion is stabilized to prevent movement in the direction of the muscle when it contacts. Additionally, muscles are able to exert greater force when lengthened than when shortened. All of the hamstring muscles, as well as the rectus femoris, are biarticular (two-joint) muscles.

As an example, the sartorius muscle becomes a better flexor at the knee when the pelvis is rotated posteriorly and stabilized by the abdominal muscles, thus increasing its total length by moving its origin farther from its insertion. This is exemplified by trying to flex the knee and cross the legs in the sitting position. One usually leans backward to flex the legs at the knees. Again, this is illustrated by kicking a football. The kicker invariably leans well backward to raise and fix the origin of the rectus femoris muscle to make it more effective as an extensor of the leg at the knee. Also, when youngsters hang by the knees, they flex the hips to fix or raise the origin of the hamstrings to make the latter more effective flexors of the knees.

The gracilis, sartorius, and semitendinosus all join together distally to form a tendinous expansion, known as the pes anserinus, which attaches to anteromedial aspect of the proximal tibia below the level of the tibial tuberosity. This attachment and the line of pull these muscles have posteromedially to the knee enable them to assist with knee flexion, particularly once the knee is flexed and the hip is externally rotated. The medial and lateral heads of the gastrocnemius attach posteriorly on the medial and lateral femoral condyles, respectively. This relationship to the knee provides the gastrocnemius a line of pull to assist with knee flexion.

Knee joint muscles—location

Muscle location closely relates to muscle function with the knee.

Anterior
   Primarily knee extension
      Rectus femoris*
      Vastus medialis
      Vastus intermedius
      Vastus lateralis
Posterior
   Primarily knee flexion
      Biceps femoris*
      Semimembranosus*
      Semitendinosus*
      Sartorius*
      Gracilis*
      Popliteus
      Gastrocnemius*

---

*Two-joint muscles; hip actions are discussed in Chapter 7 and ankle actions are discussed in Chapter 9.

Quadriceps muscles   FIG. 8.3

(kwod´ri-seps)

The ability to jump is essential in nearly all sports. Individuals who have good jumping ability always have strong quadriceps muscles that extend the leg at the knee. The quadriceps function as a decelerator when it is necessary to decrease speed for changing direction or to prevent falling when landing. This deceleration function is also evident in stopping the body when coming down from a jump. The contraction that occurs in the quadriceps during braking or decelerating actions is eccentric. This eccentric action of the quadriceps controls the slowing of movements initiated in previous phases of the sports skill.

The muscles are the rectus femoris (the only two-joint muscle of the group), vastus lateralis (the largest muscle of the group), vastus intermedius, and vastus medialis. All attach to the patella and by the patellar tendon to the tuberosity of the tibia. All are superficial and palpable, except the vastus intermedius, which is under the rectus femoris. The vertical jump is a simple test that may be used to indicate the strength or power of the quadriceps. This muscle group is generally desired to be 25% to 33% stronger than the hamstring muscle group (knee flexors).

Rectus femoris muscle

(rek´tus fem´o-ris)

Origin

Anterior inferior iliac spine of the ilium and superior margin of the acetabulum

Insertion

Superior aspect of the patella and patellar tendon to the tibial tuberosity

Action

Flexion of the hip
Extension of the knee

Palpation

Any place on the anterior surface of the femur

Innervation

Femoral nerve (L2–4)

Application, strengthening, and flexibility

When the hip is flexed, the rectus femoris becomes shorter, which reduces its effectiveness as an extensor of the knee. The work is then done primarily by the three vasti muscles.

Also see the rectus femoris discussion in Chapter 7, p. 124 (Fig. 7.9).

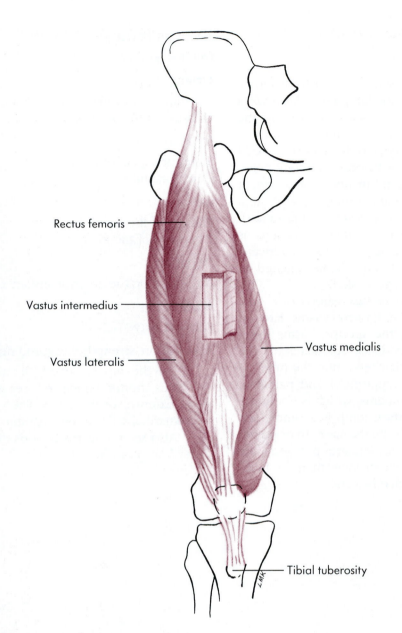

Rectus femoris

Vastus intermedius

Vastus lateralis

Vastus medialis

Tibial tuberosity

FIG. 8.3 ● Quadriceps muscle group.

Vastus lateralis (externus) muscle

FIG. 8.4

(vas´tus lat-er-a´lis)

Origin

Intertrochanteric line, anterior and inferior borders
of the greater trochanter, gluteal tuberosity, upper
half of the linea aspera and entire lateral intermuscular septum

Insertion

Lateral border of the patella and patellar tendon to
the tibial tuberosity

Action

Extension of the knee

Palpation

Anterior lateral aspect of the thigh

Innervation

Femoral nerve (L2–4)

Application, strengthening, and flexibility

All three of the vasti muscles function with the
rectus femoris in knee extension. They are typically used in walking and running and must be
used to keep the knees straight, as in standing.
The vastus lateralis has a slightly superior lateral
pull on the patella and, as a result, is occasionally
blamed in part for common lateral patellar sub-
luxation and dislocation problems.

The vastus lateralis is strengthened through knee
extension activities against resistance. Stretching
occurs by pulling the knee into maximum flexion,
such as by standing on one leg and pulling the heel
of the other leg to the buttocks.

FIG. 8.4 ● Vastus lateralis muscle, anterior view. *O*, Origin; *I*, insertion

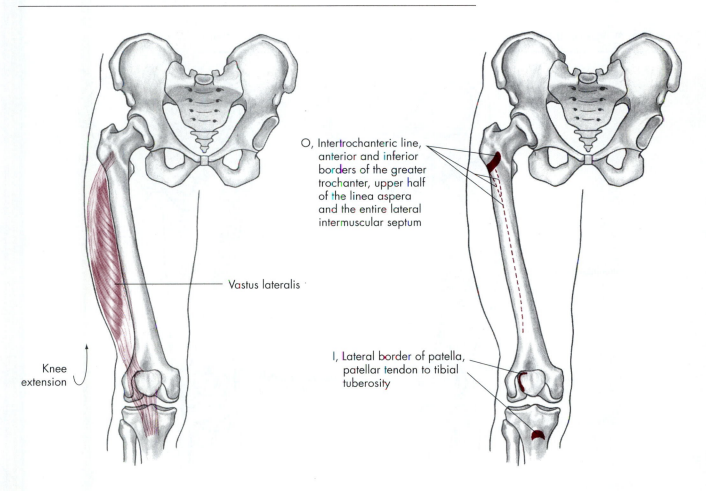

O, Intertrochanteric line,
anterior and inferior
borders of the greater
trochanter, upper half
of the linea aspera
and the entire lateral
intermuscular septum

Vastus lateralis

Knee
extension

I, Lateral border of patella,
patellar tendon to tibial
tuberosity

151

Vastus intermedius muscle   FIG. 8.5

(vas´tus in´ter-me´di-us)

Origin

Upper two-thirds of the anterior surface of the femur

Insertion

Upper border of the patella and patellar tendon to the tibial tuberosity

Action

Extension of the knee

Palpation

Cannot be palpated; lies under the rectus femoris muscle

Innervation

Femoral nerve (L2–4)

Application, strengthening, and flexibility

The three vasti muscles all contract in knee extension. They are used together with the rectus femoris in running, jumping, hopping, skipping, and walking. The vasti muscles are primarily responsible for extending the knee while the hip is flexed or being flexed. Thus, in doing a knee bend with the trunk bent forward at the hip, the vasti are exercised with little involvement of the rectus femoris. These natural activities mentioned develop the quadriceps.

Squats with a barbell of varying weights on the shoulders, depending on strength, are an excellent exercise for developing the quadriceps if done properly. Caution should be used, along with strict attention to proper technique, to avoid injuries to the knees and lower back. Leg press exercises and knee extensions with weight machines are other good exercises. Full knee flexion stretches all of the quadriceps musculature.

FIG. 8.5 ● Vastus intermedius muscle, anterior view. *O,* Origin; *I,* insertion.

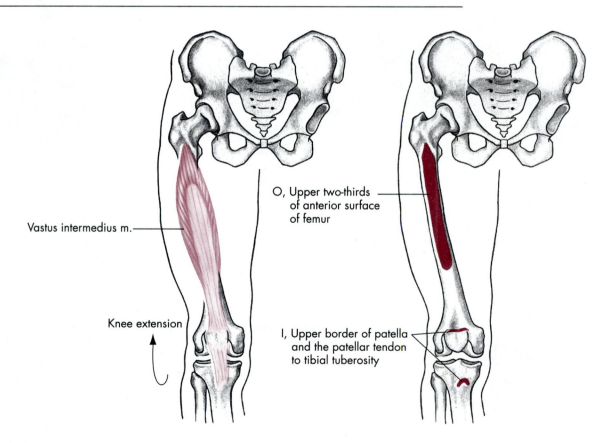

Vastus intermedius m.

Knee extension

O, Upper two-thirds of anterior surface of femur

I, Upper border of patella and the patellar tendon to tibial tuberosity

Vastus medialis (internus) muscle

FIG. 8.6

(vas´tus me-di-a´lis)

Origin

Whole length of the linea aspera and the medial condyloid ridge

Insertion

Medial half of the upper border of the patella and patellar tendon to the tibial tuberosity

Action

Extension of the knee

Palpation

Anterior medial side of the thigh near the knee joint

Innervation

Femoral nerve (L2–4)

Application, strengthening, and flexibility

The vastus medialis is thought to be very important in maintaining patellofemoral stability because of the oblique attachment of its distal fibers to the superior medial patella. This portion of the vastus medialis is referred to as the vastus medialis obliquus (VMO). The vastus medialis is strengthened similarly to the other quadriceps muscles by squats, knee extensions, and leg presses, but the VMO is not really emphasized until the last 10 to 20 degrees of knee extension. Full knee flexion stretches all of the quadriceps muscles.

FIG. 8.6 ● Vastus medialis muscle, anterior view. *O*, Origin; *I*, insertion.

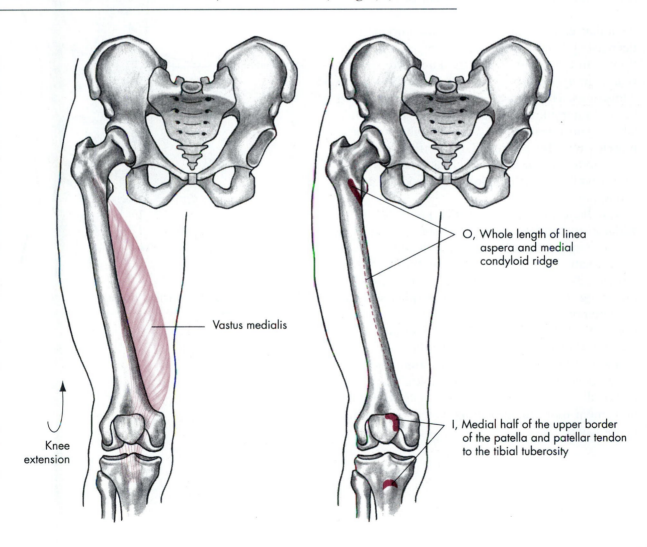

Vastus medialis

Knee extension

O, Whole length of linea aspera and medial condyloid ridge

I, Medial half of the upper border of the patella and patellar tendon to the tibial tuberosity

Hamstring muscles   FIG. 8.7

The hamstring muscle group, consisting of the biceps femoris, semimembranosus, and semitendinosus, is covered in complete detail in Chapter 7, but further discussion is included here because of its importance in knee function.

Muscle strains involving the hamstrings are very common in football and other sports that require explosive running. This muscle group is often referred to as the "running muscle" because of its function in acceleration. The hamstring muscles are antagonists to the quadriceps muscles at the knee and are named for their cordlike attachments at the knee. All of the hamstring muscles originate on the ischial tuberosity of the pelvic bone. The semitendinosus and semimembranosus insert on the anteromedial and posteromedial side of the tibia, respectively. The biceps femoris inserts on the lateral tibial condyle and head of the fibula—hence the saying "Two to the inside and one to the outside." The second head of the biceps femoris is on the linea aspera of the femur.

Special exercises to improve the strength and flexibility of this muscle group are important in decreasing knee injuries. Inability to touch the floor with the fingers when the knees are straight is largely a result of a lack of flexibility of the hamstrings. The hamstrings may be strengthened by performing knee or hamstring curls on a knee table against resistance. The flexibility of these muscles may be improved by performing slow, static stretching exercises, such as flexing the hip slowly while maintaining knee extension in a long sitting position.

The hamstrings are primarily knee flexors in addition to serving as hip extensors. Rotation of the knee can occur when it is in a flexed position. Knee rotation is brought about by the hamstring muscles. The biceps femoris externally rotates the lower leg at the knee. The semitendinosus and semimembranosus perform internal rotation. Rotation of the knee permits pivoting movements and change in direction of the body. This rotation of the knee is vital in accommodating to forces developing at the hip or ankle during directional changes in order to make the total movement more functional as well as more fluid in appearance.

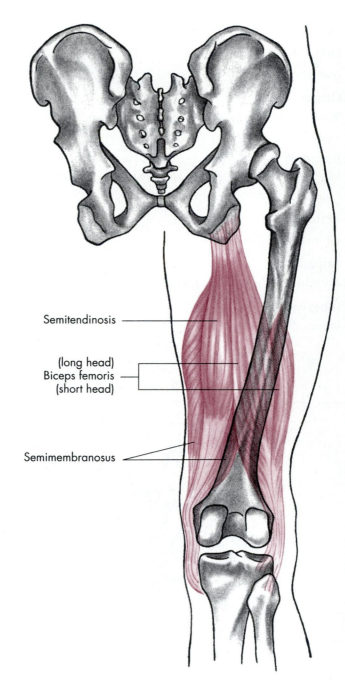

Semitendinosis

(long head)
Biceps femoris
(short head)

Semimembranosus

FIG. 8.7 ● The hamstring muscle group.

Popliteus muscle FIG. 8.8

(pop´li-te´us)

Origin

Posterior surface of the lateral condyle of the femur

Insertion

Upper posterior medial surface of the tibia

Action

Flexion of the knee
Internal rotation of the knee

Palpation

Cannot be palpated

Innervation

Tibial nerve (L5, S1)

Application, strengthening, and flexibility

The popliteus muscle is the only true flexor of the leg at the knee. All other flexors are two-joint muscles. The popliteus is vital in providing posterolateral stability to the knee. It assists the medial hamstrings in internal rotation of the lower leg at the knee.

Hanging from a bar with the legs flexed at the knee strenuously exercises the popliteus muscle. Also, the less strenuous activities of walking and running exercise this muscle. Specific efforts to strengthen this muscle combine knee internal rotation and flexion exercises against resistance. Stretching of the popliteus is difficult but may be done through passive full knee extension without flexing the hip. Passive maximum external rotation with the knee flexed approximately 20 to 30 degrees also stretches the popliteus.

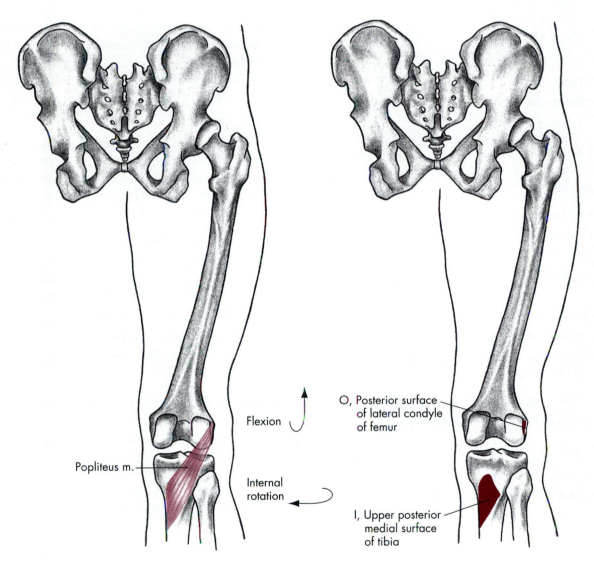

FIG. 8.8 ● Popliteus muscle, posterior view. *O*, Origin; *I*, insertion.

Web sites

Anatomy & Physiology Tutorials:

www.gwc.maricopa.edu/class/bio201/index.htm

Radiologic Anatomy Browser:

radlinux1.usuf1.usuhs.mil/rad/iong/index.html

This site has numerous radiological views of the musculoskeletal system.

University of Arkansas Medical School Gross Anatomy for Medical Students:

anatomy.uams.edu/htmlpages/anatomyhtml/gr oss.html

Dissections, anatomy tables, atlas images, links, etc.

Loyola University Medical Center: Structure of the Human Body:

www.meddean.luc.edu/lumen/MedEd/ GrossAnatomy/GA.html

An excellent site with many slides, dissections, tutorials, etc., for the study of human anatomy.

Wheeless' Textbook of Orthopaedics:

www.medmedia.com/

This site has an extensive index of links to the fractures, joints, muscles, nerves, trauma, medications, medical topics and lab tests, as well as links to orthopedic journals and other orthopedic, and medical news.

Premiere Medical Search Engine:

www.medsite.com

This site allows the reader to enter any medical condition and it will search the net to find relevant articles.

Arthroscopy.Com:

www.arthroscopy.com/sports.htm

Patient information on various musculoskeletal problems of the upper and lower extremity.

Anatomy of the Knee Tutorial:

www.ncl.ac.uk/~nccc/tutorials/knee/

Virtual Hospital:

www.vh.org

Numerous slides, patient information, etc.

Dynamic Human version 2.0 CD-ROM: The Visual Guide to Anatomy & Physiology:

www.mhhe.com/biosci/ap/dynamichuman2/

Web site that accompanies this CD-ROM.

Dynamic Human CD activities

1. Review anatomical landmarks as well as origins/insertions for the knee by clicking on **skeletal, anatomy,** and **gross anatomy,** then **pelvic girdle**.
2. Review anatomical landmarks as well as origins/insertions for the knee by clicking on **skeletal, anatomy, gross anatomy, lower limbs, femur, patella,** and then **tibia/fibula**.
3. Review each of the muscles from this chapter by clicking on **muscular, anatomy, body regions,** and then **thigh**.
4. Discuss what effects a torn lateral meniscus may have on joint range of motion. Click on **skeletal, clinical concepts,** and then **arthroscopy**.
5. Click on **skeletal, clinical concepts,** and then **MRI of knee**. Attempt to identify the ligamentous and musculoskeletal structures shown.
6. Click on **skeletal, clinical concepts,** and then **joint disorders**. Discuss what types of activities may lead to osteoarthritis. Discuss the effects that osteoarthritis may have on physical activity and vice versa.

Worksheet exercise

As an aid to learning, for in-class or out-of-class assignments, or for testing, a tear-out worksheet is found at the end of the text (p. 263).

Posterior skeletal worksheet (no. 1)

Draw and label on the worksheet the knee joint muscles.

Laboratory and review exercises

1. Locate the following parts of bones on a human skeleton and on a subject:
   a. Skeleton
      1. Head and neck of femur
      2. Greater trochanter

3. Shaft of femur
4. Lesser trochanter
5. Linea aspera
6. Adductor tubercle
7. Medial femoral condyle
8. Lateral femoral condyle
9. Patella
  b. Subject
    1. Greater trochanter
    2. Adductor tubercle
    3. Medial femoral condyle
    4. Lateral femoral condyle
    5. Patella
2. How and where can the following muscles be palpated on a human subject?
NOTE: Palpate the previously studied hip joint muscles while they are performing actions at the knee.
  a. Gracilis
  b. Sartorius
  c. Biceps femoris
  d. Semitendinosus
  e. Semimembranosus
  f. Rectus femoris
  g. Vastus lateralis
  h. Vastus intermedius
  i. Vastus medialis
  j. Popliteus
3. Be prepared to indicate on a human skeleton, by using a long rubber band, the origin and insertion of the muscles just listed.
4. Demonstrate the following movements and list the muscles primarily responsible for each.
  a. Extension of the leg at the knee
  b. Flexion of the leg at the knee
  c. Internal rotation of the leg at the knee
  d. External rotation of the leg at the knee
5. List the planes in which each of the following movements occurs. List the respective axis of rotation for each movement in each plane.
  a. Extension of the leg at the knee
  b. Flexion of the leg at the knee

c. Internal rotation of the leg at the knee
d. External rotation of the leg at the knee
6. With a laboratory partner, determine how and why maintaining the position of full knee extension limits the ability to maximally flex the hip both actively and passively. Does maintaining excessive hip flexion limit your ability to accomplish full knee extension?
7. With a laboratory partner, determine how and why maintaining the position of full knee flexion limits the ability to maximally extend the hip both actively and passively. Does maintaining excessive hip extension limit your ability to accomplish full knee flexion?
8. Compare and contrast the bony, ligamentous, articular, and cartilaginous aspects of the medial knee joint versus the lateral knee joint.
9. Research the acceptability of deep knee-bends and duck-walk activities in a physical education program and report your findings in class.
10. Prepare a report on the knee, including its ligamentous structure, joint structure, functioning, common injuries, and bracing for injuries.
11. Research preventive and rehabilitative exercises to strengthen the knee joint and report your findings in class.
12. Which muscle group about the knee would be most important to develop for an athlete with a torn anterior cruciate ligament and why? A torn posterior cruciate ligament?
13. Fill in the muscle analysis chart by listing the muscles primarily involved in each joint movement.
14. Fill in the antagonistic muscle action chart by listing the muscle(s) or parts of muscles that are antagonistic in their actions to the muscles in the left column.

Muscle analysis chart • Knee joint

Flexion	Extension
Internal rotation	External rotation

Antagonistic muscle action chart • Knee joint

Agonist	Antagonist
Biceps femoris	
Semitendinosus	
Semimembranosus	
Rectus femoris	
Vastus lateralis	
Vastus intermedius	
Vastus medialis	
Rectus femoris	
Popliteus	

References

Baker BE, et al: Review of meniscal injury and associated sports, *American Journal of Sports Medicine* 13:1, January-February 1985.

Evans W: Hamstring strength and flexibility development, *Scholastic Coach* 56:42, April 1987.

Garrick JG, Regna RK: Prophylactic knee bracing, *American Journal of Sports Medicine* 15:471, September-October 1987.

Kelly DW, et al: Patellar and quadriceps tendon ruptures—jumping knee, *American Journal of Sports Medicine* 12:375, September-October 1984.

Luttgens K, Hamilton N: *Kinesiology: scientific basis of human motion,* ed 9, Madison, WI, 1997, Brown & Benchmark.

Lysholm J, Wikland J: Injuries in runners, *American Journal of Sports Medicine* 15:168, September-October 1986.

Perreira J: Treating the quadriceps contusion, *Scholastic Coach* 57:38, October 1987.

Sieg KW, Adams SP: *Illustrated essentials of musculoskeletal anatomy,* ed 2, Gainesville, FL, 1985, Megabooks.

Stone RJ, Stone JA: *Atlas of the skeletal muscles,* 1990. McGraw-Hill Companies, Inc., New York.

Wroble RR, et al: Pattern of knee injuries in wrestling, a six-year study, *American Journal of Sports Medicine* 14:55, January-February 1986.

The ankle and foot joints

9

Objectives

- To identify on a human skeleton the most important bone features, ligaments, and arches of the ankle and foot

- To draw and label on a skeletal chart the muscles of the ankle and foot

- To demonstrate and palpate with a fellow student the movements of the ankle and foot and list their respective planes of motion and axes of rotation

- To palpate the superficial joint structures and muscles of the ankle and foot on a human subject

- To list and organize the muscles that produce movement of the ankle and foot and list their antagonists

The complexity of the foot is evidenced by the 26 bones, 19 large muscles, many small (intrinsic) muscles, and more than 100 ligaments that make up its structure.

Support and propulsion are the two functions of the foot. Proper functioning and adequate development of the muscles of the foot and practice of proper foot mechanics are essential for everyone. In our modern society, foot trouble is one of the most common ailments. Quite often, people develop poor foot mechanics or gait abnormalities secondary to improper shoe wear or other relatively minor problems. Poor foot mechanics early in life inevitably leads to foot discomfort in later years.

Walking and running may be divided into stance and swing phases. The stance phase is further divided into three components—heel-strike, midstance, and toe-off. Normally, heel-strike is characterized by landing on the heel with the foot in supination, followed immediately by the foot moving into pronation during midstance with return to supination prior to and during toe-off. The swing phase occurs when the foot leaves the ground and the leg moves forward to another point of contact. Problems often arise when the foot is too rigid and does not pronate adequately or when the foot remains in pronation past midstance. Walking differs from running in that one foot is always in contact with the ground, whereas in running there is a point when neither foot is in contact with the ground.

The fitness revolution that has occurred during the past three decades has resulted in great improvements in shoes available for sports and recreational activities. In the past, a pair of sneakers would suffice for most activities. Now there are basketball, baseball, football, jogging, soccer, tennis, walking, and cross-training shoes. Good shoes are important, but there is no substitute for adequate muscular development, strength, and proper foot mechanics.

Bones

Each foot has 26 bones, which collectively form the shape of an arch. They connect with the thigh and the remainder of the body through the fibula and tibia (Figs. 9.1 and 9.2). Body weight is transferred from the tibia to the talus and the calcaneus.

In addition to the talus and calcaneus, there are five other bones in the rear foot and midfoot known as the tarsals. Between the talus and the three cuneiform bones lies the navicular. The cuboid is located between the calcaneus and the fourth and fifth metatarsals. Distal to the tarsals are the five metatarsals, which in turn correspond to each of the five toes. The toes are known as the phalanges. There are three individual bones in each phalange, except for the great toe, which has only two. Each of these bones is known as a phalanx. Finally, there are two sesamoid bones located beneath the first metatarsophalangeal joint and contained within the flexor hallucis longus tendons.

The distal end of the tibia and fibula are enlarged and protrude horizontally and inferiorly. These bony protrusions, known as malleoli, serve as a sort of pulley for the tendons of the muscles that run directly posterior to them. This bony arrangement increases the mechanical advantage of these muscles in performing their actions of inversion and eversion. The base of the fifth metatarsal is enlarged and prominent to serve as an attachment point for the peroneus brevis and tertius. The posterior surface of the calcaneus is very prominent and serves as the attachment point for the Achilles tendon of the gastrocnemius-soleus complex.

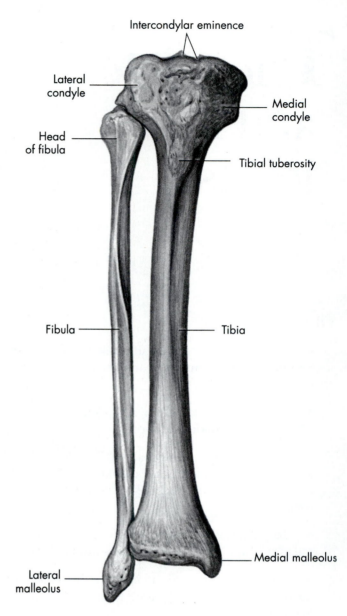

FIG. 9.1 ● Right fibula and tibia.

From Anthony CP, Kolthoff NJ: *Textbook of anatomy and physiology,* ed 9, St. Louis, 1975, Mosby.

FIG. 9.2 ● Right foot, superior view.

Modified from Anthony CP, Kolthoff NJ: *Textbook of anatomy and physiology*, ed 9, St. Loius, 1975, Mosby.

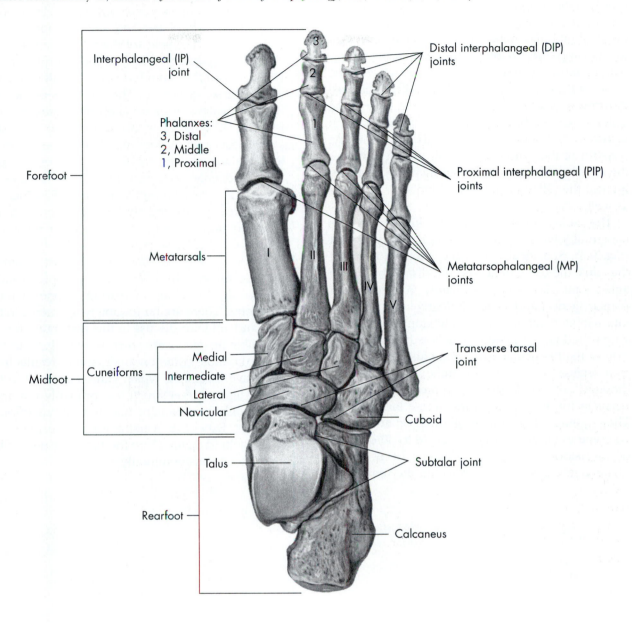

Joints

The tibia and fibula form the tibiofibular joint, a syndesmotic amphiarthrodial joint (see Fig. 9.1). The bones are joined at both the proximal and distal tibiofibular joints. In addition to the ligaments supporting both of these joints, there is a strong, dense interosseus membrane between the shafts of these two bones. Although only minimal movement is possible between these bones, the distal joint does become sprained occasionally in heavy contact sports such as football. This injury, a sprain of the syndemosis joint, usually involves the anterior inferior tibiofibular ligament and may involve the posterior inferior tibiofibular ligament as well.

The ankle joint, technically known as the talocrural joint, is a hinge or ginglymus-type joint (Fig. 9.3). Specifically, it is the joint made up of the talus, the distal tibia, and the distal fibula. The ankle joint allows approximately 50 degrees of plantar flexion and 15 to 20 degrees of dorsiflexion. Greater range of dorsiflexion is possible when the knee is flexed, which reduces the tension of the biarticular gastrocnemius muscle.

Inversion and eversion, although commonly thought to be ankle joint movements, technically occur in the subtalar and transverse tarsal joints. These joints, classified as gliding or arthrodial, combine to allow approximately 20 to 30 degrees of inversion and 5 to 15 degrees of eversion. There is minimal movement within the remainder of the intertarsal and tarsometatarsal arthrodial joints.

The phalanges join the metatarsals to form the metatarsophalangeal joints, which are classified as condyloid-type joints. The metatarsophalangeal (MP) joint of the great toe flexes 45 degrees and extends 70 degrees, whereas the interphalangeal (IP) joint can flex from 0 degrees of full extension to 90 degrees of flexion. The MP joints of the four lesser toes allow approximately 40 degrees of flexion and 40 degrees of extension. The MP joints also abduct and adduct minimally. The proximal interphalangeal (PIP) joints in the lesser toes flex from 0 degrees of extension to 35 degrees of flexion. The distal interphalangeal (DIP) joints flex 60 degrees and extend 30 degrees. There is much variation from joint to joint and from person to person in all of these joints.

Ankle sprains are one of the most common injuries among physically active people. Sprains involve the stretching or tearing of one or more ligaments. There are far too many ligaments in the foot and ankle to discuss in this text, but a few of the ankle ligaments are shown in Fig. 9.3. Far and away the most common ankle sprain results from excessive inversion, which causes damage to the lateral ligamentous structures, primarily the anterior talofibular ligament and the calcaneofibular ligament. Excessive eversion forces causing injury to the deltoid ligament on the medial aspect of the ankle occur less commonly.

FIG. 9.3 ● Right ankle joint. A, Lateral view; **B,** medial view; **C,** posterior view.

Modified from Van De Graaff KM: *Human anatomy,* ed 4, 1995, McGraw-Hill Companies Inc., New York.

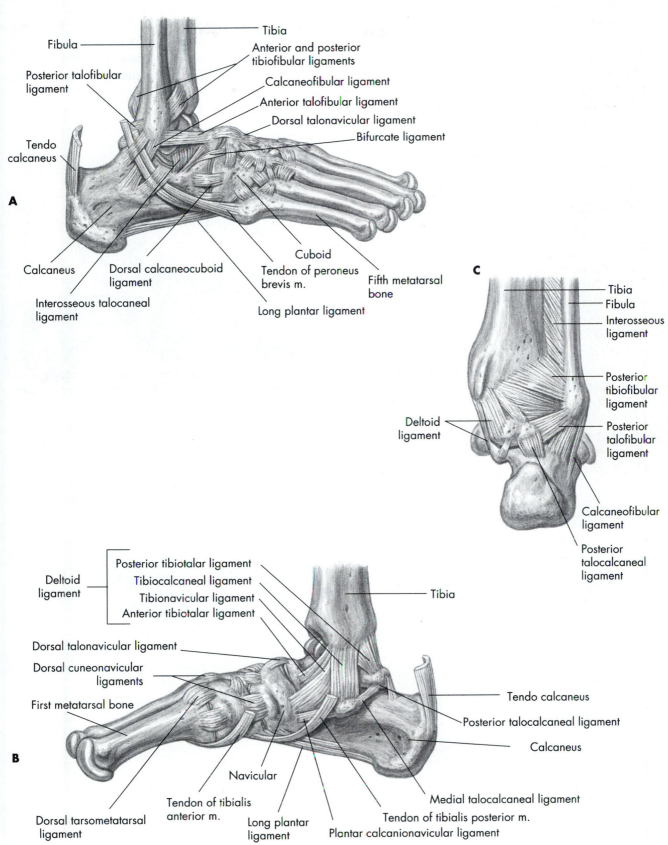

Fibula

Posterior talofibular ligament

Tendo calcaneus

A

Calcaneus

Interosseous talocaneal ligament

Dorsal calcaneocuboid ligament

Tibia

Anterior and posterior tibiofibular ligaments

Calcaneofibular ligament

Anterior talofibular ligament

Dorsal talonavicular ligament

Bifurcate ligament

Cuboid

Tendon of peroneus brevis m.

Fifth metatarsal bone

Long plantar ligament

C

Tibia

Fibula

Interosseous ligament

Posterior tibiofibular ligament

Posterior talofibular ligament

Deltoid ligament

Calcaneofibular ligament

Posterior talocalcaneal ligament

Deltoid ligament

Posterior tibiotalar ligament

Tibiocalcaneal ligament

Tibionavicular ligament

Anterior tibiotalar ligament

Dorsal talonavicular ligament

Dorsal cuneonavicular ligaments

First metatarsal bone

Tibia

Tendo calcaneus

Posterior talocalcaneal ligament

Calcaneus

Medial talocalcaneal ligament

B

Dorsal tarsometatarsal ligament

Tendon of tibialis anterior m.

Navicular

Long plantar ligament

Plantar calcanionavicular ligament

Tendon of tibialis posterior m.

163

Ligaments in the foot and the ankle maintain the position of an arch. All 26 bones in the foot are connected with ligaments. This brief discussion is focused on the longitudinal and transverse arches.

There are two longitudinal arches (Fig. 9.4). The medial longitudinal arch is located on the medial side of the foot and extends from the calcaneus bone to the talus, the navicular, the three cuneiforms, and the proximal ends of the three medial metatarsals. The lateral longitudinal arch is located on the lateral side of the foot and extends from the calcaneus to the cuboid and proximal ends of the fourth and fifth metatarsals. Individual long arches can be high, medium, or low, but a low arch is not necessarily a weak arch.

The transverse arch (see Fig. 9.4) extends across the foot from one metatarsal bone to the other.

FIG. 9.4 ● Longitudinal and transverse arches.

From Anthony CP, Kolthoff NJ: *Textbook of anatomy and physiology*, ed 9, St. Louis, 1975, Mosby.

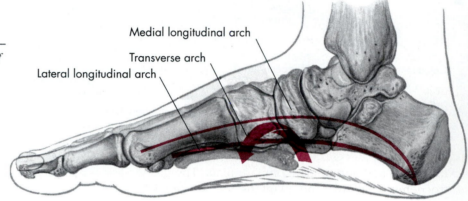

Medial longitudinal arch

Transverse arch

Lateral longitudinal arch

Ankle and foot muscles

The large number of muscles in the ankle and foot may be easier to learn if grouped according to location and function. In general, the muscles located on the anterior aspect of the ankle and foot are the dorsal flexors. Those on the posterior aspect are plantar flexors. Specifically, the gastrocnemius and the soleus are known as the triceps surae, due to their three heads collectively. Muscles that are evertors are located more to the lateral side, whereas the invertors are located medially.

The lower leg is divided into four compartments, each containing specific muscles (Fig. 9.5). Tightly surrounding and binding each compartment is a dense fascia, which facilitates venous return and prevents excessive swelling of the muscles during exercise. The anterior compartment contains the dorsiflexor group, consisting of the tibialis anterior, peroneus tertius, extensor digitorum longus, and extensor hallucis longus. The lateral compartment contains the peroneus longus and peroneus brevis—the two most powerful evertors. The posterior compartment is divided into deep and superficial compartments. The gastrocnemius, soleus, and plantaris are located in the superficial posterior compartment, while the deep posterior compartment is composed of the flexor digitorum longus, flexor hallucis longus, popliteus, and tibialis posterior. All of the muscles of the superficial posterior compartment are primarily plantar flexors. The plantaris, absent in some humans, is a vestigial biarticular muscle that contributes minimally to ankle plantar flexion. The deep posterior compartment muscles, except for the popliteus, are plantar flexors but also function as invertors.

Due to heavy demands placed on the musculature of the legs in the running activities of most sports, both acute and chronic injuries are common. "Shin splints" is a common term used to describe a painful condition of the leg that is often associated with running activities. This condition is not a specific diagnosis but, rather, is attributed to a number of specific musculotendinous injuries. Most often the tibialis posterior, medial soleus, or anterior tibialis is involved, but the extensor digitorum longus may also be involved. Shin splints may be prevented in part by stretching the plantar flexors and strengthening the dorsiflexors.

Additionally, painful cramps caused by acute muscle spasm in the gastrocnemius and soleus occur somewhat commonly and may be relieved through active and passive dorsiflexion. Also, a very disabling injury involves the complete rupture of the strong Achilles tendon, which connects these two plantar flexors to the calcaneus.

NOTE: A number of the ankle and foot muscles are capable of helping produce more than one movement.

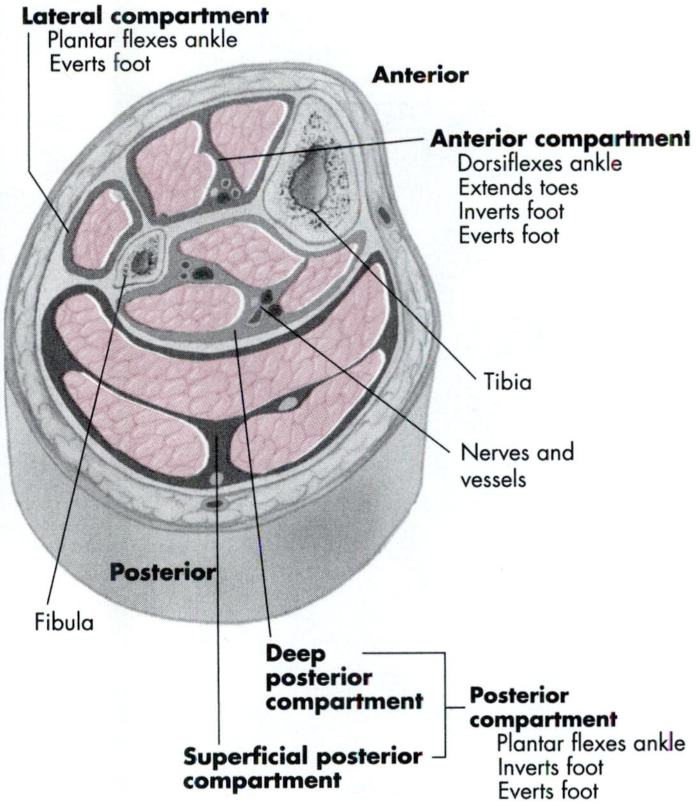

Lateral compartment
Plantar flexes ankle
Everts foot

Anterior

Anterior compartment
Dorsiflexes ankle
Extends toes
Inverts foot
Everts foot

Tibia

Nerves and vessels

Posterior

Fibula

Deep posterior compartment

Superficial posterior compartment

Posterior compartment
Plantar flexes ankle
Inverts foot
Everts foot

FIG. 9.5 • Cross section of the leg, demonstrating the muscular compartments.

From Seeley RR, Stephens TD, Tate P: *Anatomy and physiology,* ed 3, St. Louis, 1995, Mosby.

Ankle and foot muscles

Plantar flexors
- Gastrocnemius
- Flexor digitorum longus
- Flexor hallucis longus
- Peroneus longus
- Peroneus brevis
- Plantaris
- Soleus
- Tibialis posterior

Evertors
- Peroneus longus
- Peroneus brevis
- Peroneus tertius
- Extensor digitorum longus

Dorsiflexors
- Tibialis anterior
- Peroneus tertius
- Extensor digitorum longus (extensor of the lesser toes)
- Extensor hallucis longus (extensor of the great toe)

Invertors
- Tibialis anterior
- Tibialis posterior
- Flexor digitorum longus (flexor of the lesser toes)
- Flexor hallucis longus (flexor of the great toe)

FIG. 9.6 ● Movements of the ankle and foot.

Movements   FIG. 9.6

Dorsiflexion (flexion): dorsal flexion; movement of the top of the ankle and foot toward the anterior tibia bone

Plantar flexion (extension): movement of the ankle and foot away from the tibia

Eversion: turning the ankle and foot outward, abduction, away from the midline; weight is on the medial edge of the foot

Inversion: turning the ankle and foot inward, adduction, toward the midline; weight is on the lateral edge of the foot

Toe flexion: movement of the toes toward the plantar surface of the foot

Toe extension: movement of the toes away from the plantar surface of the foot

Pronation: a combination of ankle dorsiflexion, subtalar eversion, and forefoot abduction (toe-out)

Supination: a combination of ankle plantar flexion, subtalar inversion, and forefoot adduction (toe-in)

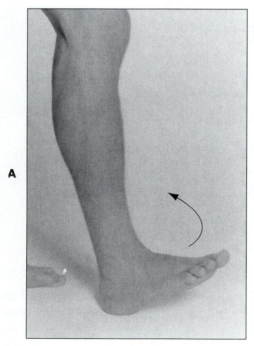

Dorsiflexion

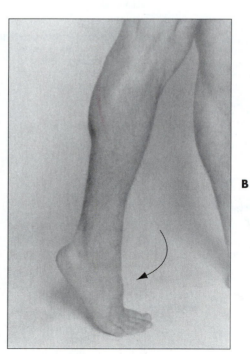

Plantar flexion

FIG. 9.6 continued ● Movements of the ankle and foot.

C

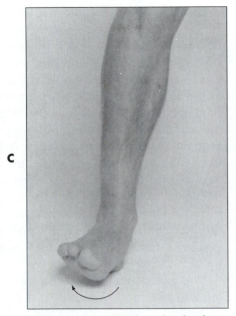

D

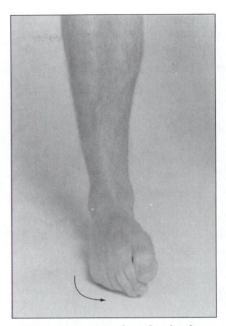

**Transverse tarsal and subtalar
eversion**

**Transverse tarsal and subtalar
inversion**

E

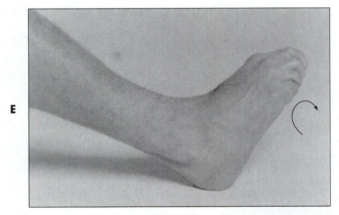

Flexion of the toes

F

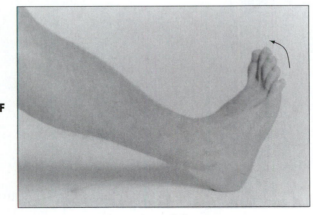

Extension of the toes

Gastrocnemius muscle   FIG. 9.7

(gas-trok-neˊmi-us)

Origin

Medial head: posterior surface of the medial femoral
  condyle
Lateral head: posterior surface of the lateral femoral
  condyle

Insertion

Posterior surface of the calcaneus (Achilles tendon)

Action

Plantar flexion of the ankle
Flexion of the knee

Palpation

Easiest muscle in the lower extremity to palpate;
  upper posterior aspect of the lower leg

Innervation

Tibial nerve (S1, 2)

Application, strengthening, and flexibility

Because the gastrocnemius is a biarticular muscle, it is more effective as a knee flexor if the ankle is dorsiflexed and more effective as a plantar flexor of the foot if the knee is held in extension. This is observed when one sits too close to the wheel in driving a car, which significantly shortens the entire muscle, reducing its effectiveness. When the knees are bent, the muscle becomes an ineffective plantar flexor, and it is more difficult to depress the brakes. Running, jumping, hopping, and skipping exercises all depend significantly on the gastrocnemius and soleus to propel the body upward and forward. Heel-raising exercises with the knees in full extension and the toes resting on a block of wood are an excellent way to strengthen the muscle through the full range of motion. By holding a barbell on the shoulders, the resistance may be increased.

The gastrocnemius may be stretched by standing and placing both palms on a wall about 3 feet away and leaning into the wall. The feet should be pointed straight ahead, and the heels should remain on the floor. The knees should remain fully extended throughout the exercise to accentuate the stretch on the gastrocnemius.

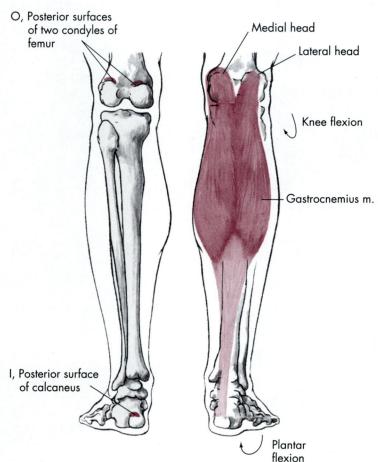

O, Posterior surfaces
  of two condyles of
  femur

Medial head

Lateral head

Knee flexion

Gastrocnemius m.

I, Posterior surface
  of calcaneus

Plantar
flexion

FIG. 9.7 ● Gastrocnemius muscle, posterior view. *O*, Origin; *I*, insertion.

Soleus muscles   FIG. 9.8

(so´le-us)

Origin

Posterior surface of the proximal fibula and proximal two-thirds of the posterior tibial surface

Insertion

Posterior surface of the calcaneus (Achilles tendon)

Action

Plantar flexion of the ankle

Palpation

Posteriorly under the gastrocnemius muscle on the lateral side of the lower leg

Innervation

Tibial nerve (S1, 2)

Application, strengthening, and flexibility

The soleus muscle is one of the most important plantar flexors of the ankle. Some anatomists believe that it is nearly as important in this movement as the gastrocnemius. This is especially true when the knee is flexed. When one rises up on the toes, the soleus muscle can plainly be seen on the outside of the lower leg if one has exercised the legs extensively, as in running and walking.

The soleus muscle is used whenever the ankle plantar flexes. Any movement with body weight on the foot with the knee flexed or extended calls it into action. When the knee is flexed slightly, the effect of the gastrocnemius is reduced, thereby placing more work on the soleus. Running, jumping, hopping, skipping, and dancing on the toes are all exercises that depend heavily on the soleus. It may be strengthened through any plantar flexion exercise against resistance, particularly if the knee is flexed slightly to deemphasize the gastrocnemius. Heel-raising exercises as described for the gastrocnemius, except with the knees flexed slightly, are one way to isolate this muscle for strengthening. Resistance may be increased by holding a barbell on the shoulders.

The soleus is stretched in the same manner as for the gastrocnemius, except that the knees must be flexed slightly, which releases the stretch on the gastrocnemius and places it on the soleus. Again, it is important to attempt to keep the heels on the floor.

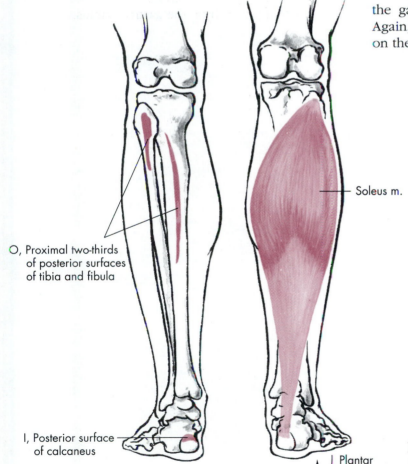

Soleus m.

O, Proximal two-thirds of posterior surfaces of tibia and fibula

I, Posterior surface of calcaneus

Plantar flexion

FIG. 9.8 ● Soleus muscle, posterior view. O, Origin; I, insertion.

Peroneus longus muscle   FIG. 9.9

(per-o-ne´us long´gus)

Origin

Head and upper two-thirds of the lateral surface of the fibula

Insertion

Undersurfaces of the medial cuneiform and first metatarsal bones

Action

Eversion of the foot
Plantar flexion of the ankle

Palpation

Upper lateral side of the tibia; just posterolateral from the tibialis anterior and extensor digitorum longus

Innervation

Superficial peroneal nerve (L4–5, S1)

Application, strengthening, and flexibility

The peroneus longus muscle passes posteroinferiorly to the lateral malleolus and under the foot from the outside to under the inner surface. Because of its line of pull, it is a strong evertor and assists in plantar flexion.

When the peroneus longus muscle is used effectively with the other ankle flexors, it helps bind the transverse arch as it flexes. Developed without the other plantar flexors, it would produce a weak, everted foot. In running, jumping, hopping, and skipping, the foot should be placed so that it is pointing forward to ensure proper development of the group. Walking barefoot or in stocking feet on the inside of the foot (everted position) is the best exercise for this muscle.

Eversion exercises to strengthen this muscle may be performed by turning the sole of the foot outward while resistance is applied in the opposite direction.

The peroneus longus may be stretched by passively taking the foot into extreme inversion and dorsiflexion while the knee is flexed.

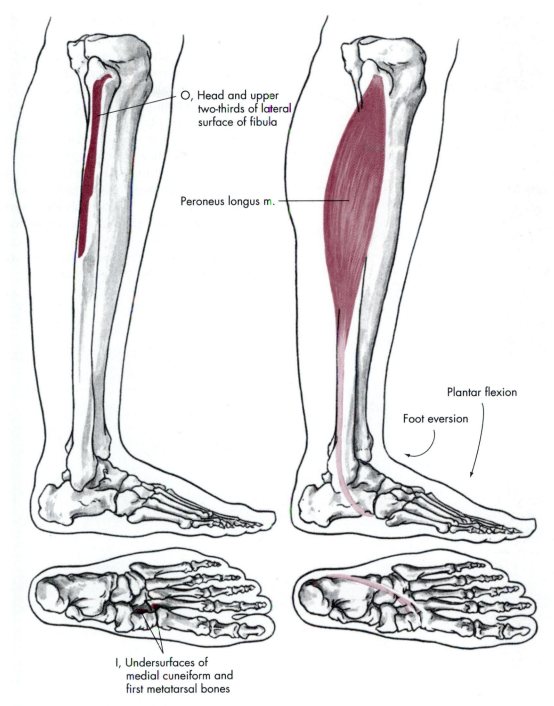

O, Head and upper two-thirds of lateral surface of fibula

Peroneus longus m.

Plantar flexion

Foot eversion

I, Undersurfaces of medial cuneiform and first metatarsal bones

FIG. 9.9 ● Peroneus longus muscle, lateral & plantar views. *O*, Origin; *I*, insertion.

Peroneus brevis muscle   FIG. 9.10

(per-o-ne´us bre´vis)

Origin

Lower two-thirds of the lateral surface of the fibula

Insertion

Tuberosity of the fifth metatarsal bone

Action

Eversion of the foot

Plantar flexion of the ankle

Palpation

Tendon of the muscle at the proximal end of the
  fifth metatarsal

Innervation

Superficial peroneal nerve (L4–5, S1)

Application, strengthening, and flexibility

The peroneus brevis muscle passes posteroinferiorly to the lateral malleolus to pull on the base of the fifth metatarsal. It is a primary evertor of the foot and assists in plantar flexion. In addition, it aids in maintaining the longitudinal arch as it depresses the foot.

The peroneus brevis muscle is exercised with other plantar flexors in the powerful movements of running, jumping, hopping, and skipping. It may be strengthened in a fashion similar to the peroneus longus by performing eversion exercises, such as turning the sole of the foot outward against resistance.

The peroneus brevis is stretched in the same manner as the peroneus longus.

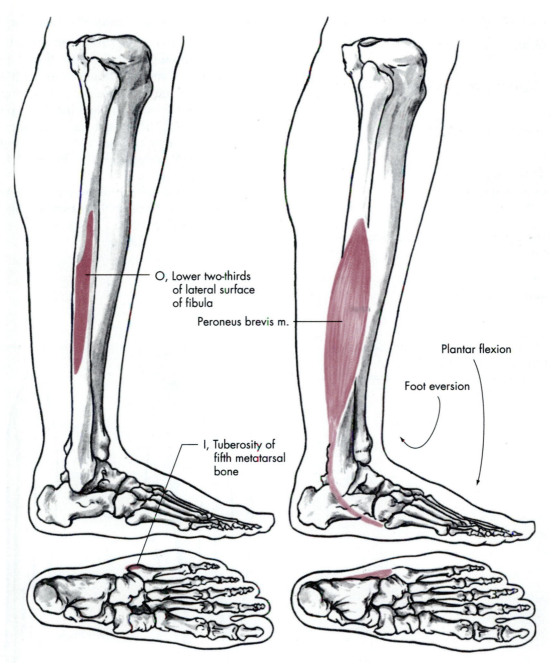

O, Lower two-thirds
of lateral surface
of fibula

Peroneus brevis m.

Plantar flexion

Foot eversion

I, Tuberosity of
fifth metatarsal
bone

FIG. 9.10 ● Peroneus brevis muscle, lateral & plantar views. *O*, Origin; *I*, insertion.

Peroneus tertius muscle   FIG. 9.11

(per-o-ne´us ter´shi-us)

Origin

Distal third of the anterior fibula

Insertion

Base of the fifth metatarsal

Action

Eversion of the foot
Dorsal flexion of the ankle

Palpation

Lateral to the extensor digitorum longus tendon on
   the anterolateral aspect of the foot

Innervation

Deep peroneal nerve (L4–5, S1)

FIG. 9.11 ● Peroneus tertius muscle, anterior
view. *O*, Origin; *I*, insertion.

Application, strengthening, and flexibility

The peroneus tertius, absent in some humans, is a
small muscle that assists in dorsal flexion and
eversion. Some authorities refer to it as the fifth
tendon of the extensor digitorum longus. It may
be strengthened by pulling the foot up toward the
shin against a weight or resistance. Everting
the foot against resistance, such as weighted eversion towel drags, can also be used for strength
development.

The peroneus tertius may be stretched by passively taking the foot into extreme inversion and
plantar flexion.

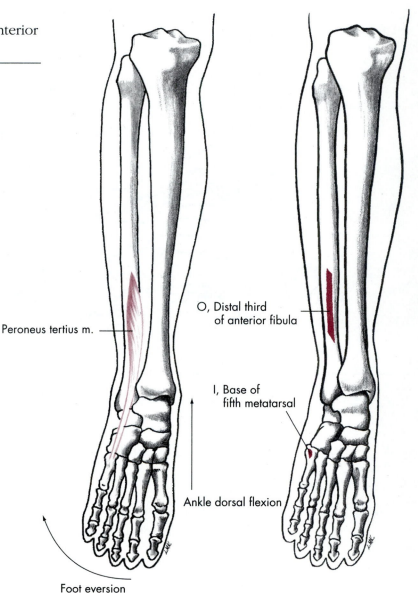

Peroneus tertius m.

O, Distal third
of anterior fibula

I, Base of
fifth metatarsal

Ankle dorsal flexion

Foot eversion

Extensor digitorum longus muscle

FIG. 9.12

(eks-ten´sor dij-i-to´rum long´gus)

Origin

Lateral condyle of the tibia, head of the fibula, and
upper two-thirds of the anterior surface of the
fibula

Insertion

Tops of the middle and distal phalanges of the four
lesser toes

Action

Extension of the four lesser toes
Dorsal flexion of the ankle
Eversion of the foot

Palpation

Second muscle on the lateral side of the tibia; upper
lateral side of the tibia; just posterolateral to the
tibialis anterior

Innervation

Deep peroneal nerve (L4–5, S1)

Application, strengthening, and flexibility

Strength is necessary in the extensor digitorum
longus muscle to maintain balance between the
plantar and the dorsal flexors.

Action that involves dorsal flexion of the ankle
and extension of the toes against resistance
strengthens both the extensor digitorum longus
and the extensor hallucis longus muscles. This
may be accomplished by manually applying a
downward force on the toes while attempting to
extend them up.

The extensor digitorum longus may be
stretched by passively taking the four lesser toes
into full flexion while the foot is inverted and
plantar flexed.

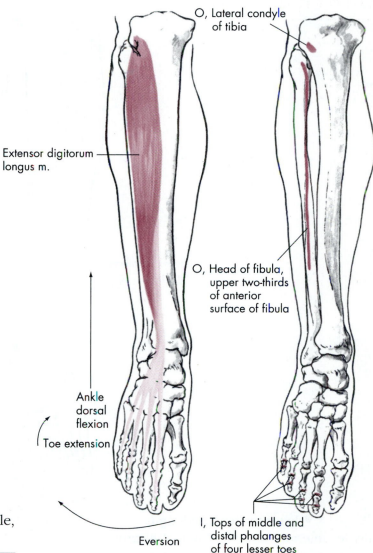

O, Lateral condyle
of tibia

Extensor digitorum
longus m.

O, Head of fibula,
upper two-thirds
of anterior
surface of fibula

Ankle
dorsal
flexion

Toe extension

Eversion

I, Tops of middle and
distal phalanges
of four lesser toes

FIG. 9.12 ● Extensor digitorum longus muscle,
anterior view. O, Origin; I, insertion.

Extensor hallucis longus muscle

FIG. 9.13

(eks-ten´sor hal-u´sis long´gus)

Origin

Middle two-thirds of the medial surface of the anterior fibula

Insertion

Base of the distal phalanx of the great toe

Action

Dorsiflexion of the ankle
Extension of the great toe
Weak inversion of the foot

Palpation

Near the great toe on the dorsal surface

Innervation

Deep peroneal nerve (L4–5, S1)

Application, strengthening, and flexibility

The three dorsiflexors of the foot—tibialis anterior, extensor digitorum longus, and extensor hallucis longus—may be exercised by attempting to walk on the heels with the ankle flexed dorsally and toes extended. Extension of the great toe, as well as ankle dorsiflexion against resistance, will provide strengthening for this muscle.

The extensor hallucis longus may be stretched by passively taking the great toe into full flexion while the foot is everted and plantar flexed.

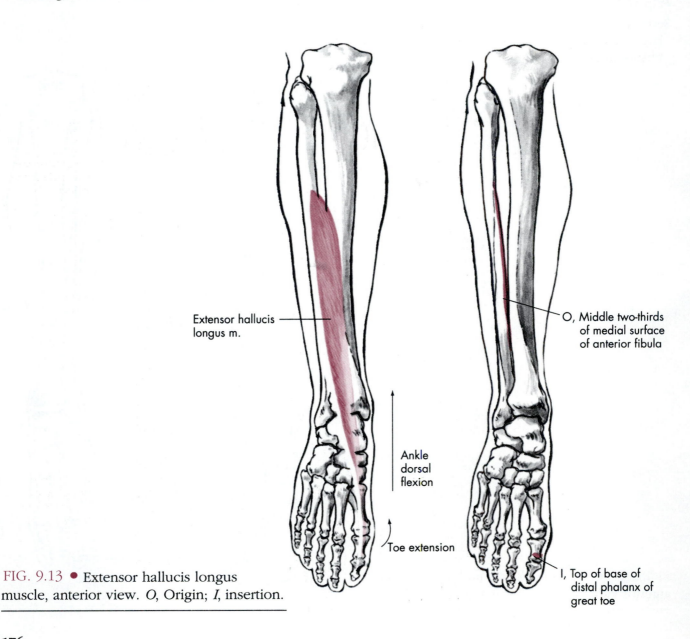

Extensor hallucis longus m.

O, Middle two-thirds of medial surface of anterior fibula

Ankle dorsal flexion

Toe extension

I, Top of base of distal phalanx of great toe

FIG. 9.13 ● Extensor hallucis longus muscle, anterior view. *O*, Origin; *I*, insertion.

Tibialis anterior muscle   FIG. 9.14

(tib-i-a´lis ant-te´ri-or)

Origin

Upper two-thirds of the lateral surface of the tibia

Insertion

Inner surface of the medial cuneiform and the first
  metatarsal bone

Action

Dorsal flexion of the ankle
Inversion of the foot

Palpation

First muscle to the lateral side of the tibia

Innervation

Deep peroneal nerve (L4–5, S1)

Application, strengthening, and flexibility

By its insertion, the tibialis anterior muscle is in a
fine position to hold up the inner margin of the
foot. However, as it contracts concentrically, it
dorsiflexes the ankle and is used as an antagonist
to the plantar flexors of the ankle. The tibialis anterior is forced to contract strongly when a person
ice skates or walks on the outside of the foot. It
strongly supports the long arch in inversion.

Walking barefoot or in stocking feet on the outside of the foot (inversion) is an excellent exercise
for the tibialis anterior muscle.

Turning the sole of the foot to the inside
against resistance to perform inversion exercises
is one way to strengthen this muscle. Dorsal flexion exercises against resistance may also be used
for this purpose.

The tibialis anterior may be stretched by passively taking the foot into extreme eversion and
plantar flexion.

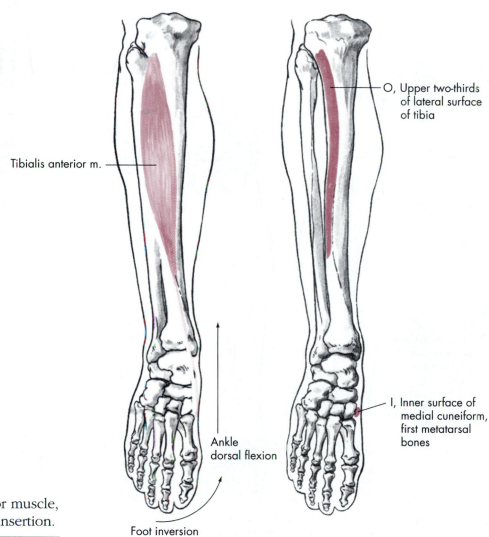

Tibialis anterior m.

O, Upper two-thirds
of lateral surface
of tibia

I, Inner surface of
medial cuneiform,
first metatarsal
bones

Ankle
dorsal flexion

Foot inversion

FIG. 9.14 ● Tibialis anterior muscle,
anterior view. *O*, Origin; *I*, insertion.

Tibialis posterior muscle   FIG. 9.15

(tib-i-a´lis pos-te´ri-or)

Origin

Posterior surface of the upper half of the interosseus membrane and adjacent surfaces of the tibia and fibula

Insertion

Lower inner surfaces of the navicular and cuneiform bones and bases of the second, third, fourth, and fifth metatarsal bones

Action

Plantar flexion of the ankle
Inversion of the foot

Palpation

Cannot be palpated

Innervation

Tibial nerve (L5, S1)

Application, strengthening, and flexibility

Passing down the back of the leg, under the medial malleolus, then forward to the navicular and medial cuneiform bones, the tibialis posterior muscle pulls down from the underside and, when contracted concentrically, inverts and plantar flexes the foot. "Shin splints" is a slang term frequently used to describe an often chronic condition in which the tibialis posterior, tibialis anterior, and extensor digitorum longus muscles are inflamed. This inflammation is usually a tendonitis of one or more of these structures but may be a result of stress fracture, periostitis, tibial stress syndrome, or a compartment syndrome. Sprints and long-distance running are common causes, particularly if the athlete has not developed appropriate strength, flexibility, and endurance in the lower leg musculature.

Use of the tibialis posterior muscle in plantar flexion and inversion gives support to the longitudinal arch of the foot. This muscle is generally strengthened by performing heel raises, as described for the gastrocnemius and soleus, as well as inversion exercises against resistance.

The tibialis posterior may be stretched by passively taking the foot into extreme eversion and dorsiflexion while the knee is flexed.

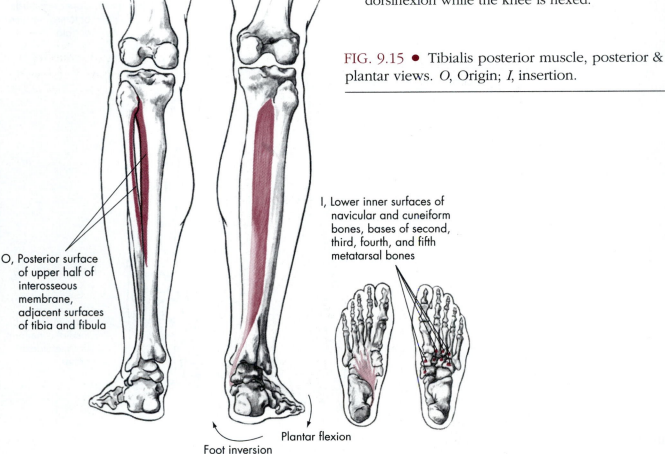

FIG. 9.15 ● Tibialis posterior muscle, posterior & plantar views. *O*, Origin; *I*, insertion.

O, Posterior surface of upper half of interosseous membrane, adjacent surfaces of tibia and fibula

I, Lower inner surfaces of navicular and cuneiform bones, bases of second, third, fourth, and fifth metatarsal bones

Plantar flexion

Foot inversion

Flexor digitorum longus muscle

FIG. 9.16

(fleks´or dij-i-to´rum long´gus)

Origin

Middle third of the posterior surface of the tibia

Insertion

Base of the distal phalanx of each of the four lesser toes

Action

Flexion of the four lesser toes
Plantar flexion of the ankle
Inversion of the foot

Palpation

Cannot be palpated

Innervation

Tibial nerve (L5, S1)

Application, strengthening, and flexibility

Passing down the back of the lower leg under the medial malleolus and then forward, the flexor digitorum longus muscle draws the four lesser toes down into flexion toward the heel as it plantar flexes the ankle. It is very important in helping other foot muscles maintain the longitudinal arch. Walking, running, and jumping do not necessarily call the flexor digitorum longus muscle into action. Some of the weak foot and ankle conditions result from ineffective use of the flexor digitorum longus. Walking barefoot with the toes curled downward toward the heels and with the foot inverted will exercise this muscle. It may be strengthened by performing towel grabs against resistance in which the heel rests on the floor while the toes extend to grab a flat towel and then flex to pull the towel under the foot. This may be repeated numerous times, with a small weight placed on the opposite end of the towel for added resistance.

The flexor digitorum longus may be stretched by passively taking the four lesser toes into extreme extension while the foot is everted and dorsiflexed. The knee should be flexed.

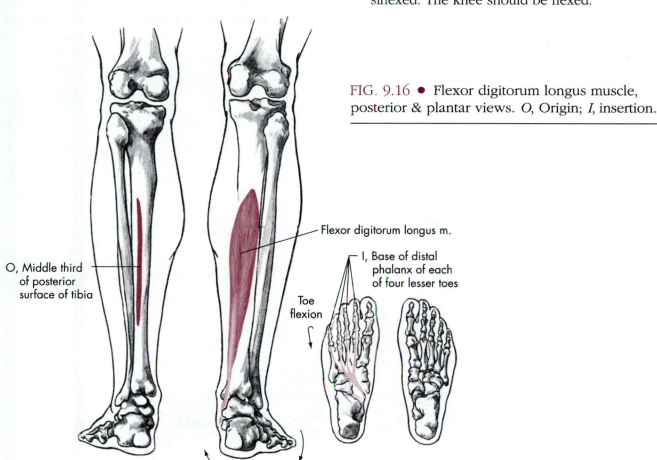

FIG. 9.16 ● Flexor digitorum longus muscle, posterior & plantar views. *O*, Origin; *I*, insertion.

O, Middle third of posterior surface of tibia

Flexor digitorum longus m.

I, Base of distal phalanx of each of four lesser toes

Toe flexion

Foot inversion

Plantar flexion

Flexor hallucis longus muscle

FIG. 9.17

(fleks´or hal-u´sis long´gus)

Origin

Middle two-thirds of the posterior surface of the
  fibula

Insertion

Base of the distal phalanx of the big toe, under the
  surface

Action

Flexion of the great toe
Inversion of the foot
Plantar flexion of the ankle

Palpation

Anteromedial to the Achilles tendon near the heel

Innervation

Tibial nerve (L5, S1–2)

Application, strengthening, and flexibility

Pulling from the underside of the great toe, the
flexor hallucis longus muscle may work independently of the flexor digitorum longus muscle or

with it. If these two muscles are poorly developed,
they cramp easily when they are called on to do
activities to which they are unaccustomed.

These muscles are used effectively in walking if
the toes are used (as they should be) in maintaining balance as each step is taken. Walking "with
the toes" rather than "over" them is an important
action for them.

When the gastrocnemius, soleus, tibialis posterior, peroneus longus, peroneus brevis, flexor digitorum longus, flexor digitorum brevis, and flexor
hallucis longus muscles are all used effectively in
walking, the strength of the ankle is evident. If an
ankle and a foot are weak, in most cases it is because of lack of use of all the muscles just mentioned. Running, walking, jumping, hopping, and
skipping provide exercise for this muscle group.
The flexor hallucis longus muscle may be specifically strengthened by performing towel grabs as
described for the flexor digitorum longus.

The flexor hallucis longus may be stretched by
passively taking the great toe into extreme extension while the foot is everted and dorsiflexed. The
knee should be flexed.

FIG. 9.17 ● Flexor hallucis longus muscle,
medial view. *O*, Origin; *I*, insertion.

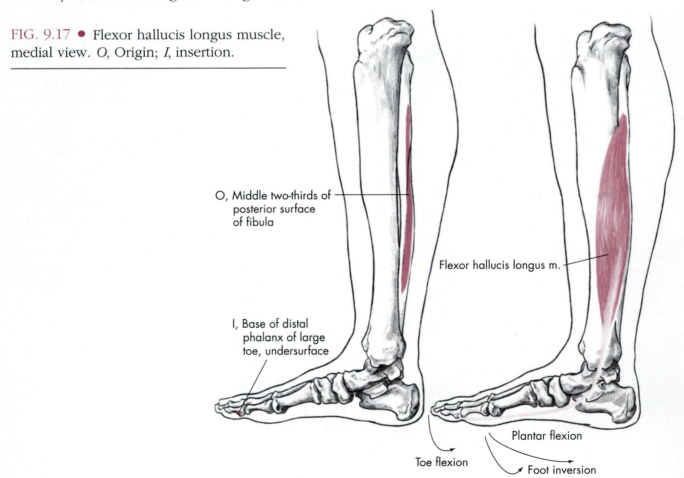

O, Middle two-thirds of
posterior surface
of fibula

Flexor hallucis longus m.

I, Base of distal
phalanx of large
toe, undersurface

Plantar flexion

Toe flexion

Foot inversion

Intrinsic muscles of the foot   FIG. 9.18

The intrinsic muscles of the foot have their origins and insertions on the bones within the foot (Fig. 9.18). One of these muscles, the extensor digitorum brevis, is found on the dorsum of the foot. The remainder are found in a plantar compartment in four layers on the plantar surface of the foot. The following muscles are found in the four layers:

First (superficial) layer: abductor hallucis, flexor digitorum brevis, abductor digiti minimi (quinti)
Second layer: quadratus plantae, lumbricales (four)
Third layer: flexor hallucis brevis, adductor hallucis, flexor digiti minimi (quinti) brevis
Fourth (deep) layer: dorsal interossei (four), plantar interossei (three)

The intrinsic foot muscles may be grouped by location as well as by the parts of the foot they act. The abductor hallucis, flexor hallucis brevis, and adductor hallucis all insert either medially or laterally on the proximal phalanx of the great toe. The abductor hallucis and flexor hallucis brevis are located somewhat medially, whereas the adductor hallucis is more centrally located beneath the metatarsals.

The quadratus plantae, four lumbricales, four dorsal interossei, three plantar interossei, flexor digitorum brevis, and extensor digitorum brevis are all located somewhat centrally. All are beneath the foot except the extensor digitorum brevis, which is the only intrinsic muscle in the foot located in the dorsal compartment. Although the entire extensor digitorum brevis has its origin on the anterior and lateral calcaneus, some anatomists refer to its first tendon as the extensor hallucis brevis in order to maintain consistency in naming according to function and location.

Located laterally beneath the foot are the abductor digiti minimi and the flexor digiti minimi brevis, which both insert on the lateral aspect of the base of the proximal phalanx of the fifth phalange. Because of these two muscles' insertion and action on the fifth toe, the name "quinti" is sometimes used instead of minimi.

Four muscles act on the great toe. The abductor hallucis is solely responsible for abduction of the great toe but assists the flexor hallucis brevis in flexing the great toe at the metatarsophalangeal joint. The adductor hallucis is the sole adductor of the great toe, while the extensor digitorum brevis is the only intrinsic extensor of the great toe at the metatarsophalangeal joint.

The four lumbricales are flexors of the second, third, fourth, and fifth phalanges at their metatarsophalangeal joints, while the quadratus plantae are flexors of these phalanges at their distal interphalangeal joints. The three plantar interossei are adductors and flexors of the proximal phalanxes of the third, fourth, and fifth phalanges, while the four dorsal interossei are abductors and flexors of the second, third, and fourth phalanges, also at their metatarsophalangeal joints. The flexor digitorum brevis flexes the middle phalanxes of the second, third, fourth, and fifth phalanges. The extensor digitorum brevis, as previously mentioned, is an extensor of the great toe but also extends the second, third, and fourth phalanges at their metatarsophalangeal joints.

There are two muscles that act solely on the fifth toe. The proximal phalanx of the fifth phalange is abducted by the abductor digiti minimi and is flexed by the flexor digiti minimi brevis.

Refer to Table 9.1 for further details regarding the intrinsic muscles of the foot.

Muscles are developed and maintain their strength only when they are used. One factor in the great increase in weak foot conditions is the lack of exercise to develop these muscles. Walking is one of the best activities for maintaining and developing the many small muscles that help support the arch of the foot. Some authorities advocate walking without shoes or with some of the newer shoes designed to enhance proper mechanics. Additionally, towel exercises such as those described for the flexor digitorum longus and flexor hallucis longus are helpful in strengthening the intrinsic muscles of the foot.

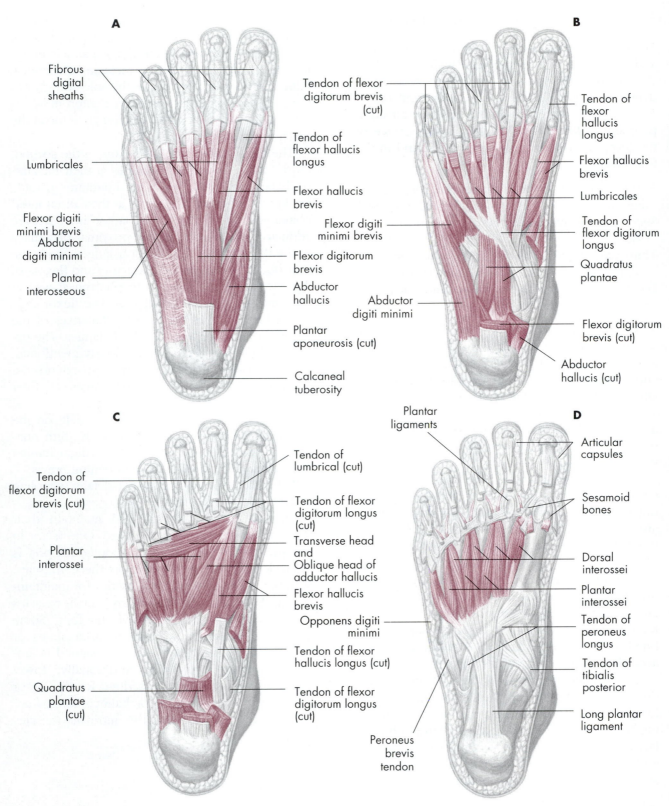

A

Fibrous digital sheaths

Lumbricales

Flexor digiti minimi brevis
Abductor digiti minimi

Plantar interosseous

Tendon of flexor hallucis longus

Flexor hallucis brevis

Flexor digitorum brevis

Abductor hallucis

Plantar aponeurosis (cut)

Calcaneal tuberosity

B

Tendon of flexor digitorum brevis (cut)

Tendon of flexor hallucis longus

Flexor hallucis brevis

Lumbricales

Tendon of flexor digitorum longus

Quadratus plantae

Flexor digitorum brevis (cut)

Abductor hallucis (cut)

Flexor digiti minimi brevis

Abductor digiti minimi

C

Tendon of flexor digitorum brevis (cut)

Plantar interossei

Quadratus plantae (cut)

Tendon of lumbrical (cut)

Tendon of flexor digitorum longus (cut)

Transverse head and
Oblique head of adductor hallucis

Flexor hallucis brevis

Opponens digiti minimi

Tendon of flexor hallucis longus (cut)

Tendon of flexor digitorum longus (cut)

Plantar ligaments

Peroneus brevis tendon

D

Articular capsules

Sesamoid bones

Dorsal interossei

Plantar interossei

Tendon of peroneus longus

Tendon of tibialis posterior

Long plantar ligament

FIG. 9.18 ● The four musculotendinous layers of the plantar aspect of the foot, detailing the intrinsic muscles. **A**, Superficial layer; **B**, second layer; **C**, third layer; **D**, deep layer.

Modified from Van De Graaff KM: *Human anatomy*, ed 4, 1995, McGraw-Hill Companies, Inc., New York.

TABLE 9.1 • Intrinsic muscles of the foot

Muscle	Origin	Insertion	Action	Palpation	Innervation
Abductor hallucis	Tuberosity of calcaneus, flexor retinaculum, plantar aponeurosis	Medial aspect of base of 1st proximal phalanx	MP flexion, abduction of 1st phalanx	Cannot be palpated	Medial plantar nerve (L4, 5)
Flexor hallucis brevis	Cuboid, lateral cuneiform	Medial head: medial aspect of 1st proximal phalanx Lateral head: lateral aspect of 1st proximal phalanx	MP flexion of 1st phalange	Cannot be palpated	Medial plantar nerve (L4, 5, S1)
Adductor hallucis	Oblique head: 2nd, 3rd, and 4th metatarsals and sheath of peroneus longus tendon Transverse head: plantar metatarsophalangeal ligaments of 3rd, 4th, and 5th phalanges and transverse metatarsal ligaments	Lateral aspect of base of 1st proximal phalanx	MP adduction of 1st phalange	Cannot be palpated	Lateral plantar nerve (S1, 2)
Quadratus plantae	Medial head: medial surface of calcaneus Lateral head: lateral border of inferior surface of calcaneus	Lateral margin of flexor digitorum longus tendon	DIP flexion of 2nd, 3rd, 4th, and 5th phalanges	Cannot be palpated	Lateral plantar nerve (S1, 2)
Lumbricales (4)	Tendons of flexor digitorum longus	Dorsal surface of 2nd, 3rd, 4th, and 5th proximal phalanxes	MP flexion of 2nd, 3rd, 4th, and 5th phalanges	Cannot be palpated	1st lumbricales: medial plantar nerve (L4, 5) 2nd, 3rd, 4th lumbricales: lateral plantar nerve (S1, 2)
Dorsal interossei (4)	Two heads on shafts of adjacent metatarsals	1st interossei: medial aspect of 2nd proximal phalanx; 2nd, 3rd, and 4th interossei: lateral aspect of 2nd, 3rd, and 4th proximal phalanxes	MP abduction and flexion of 2nd, 3rd, and 4th phalanges	Cannot be palpated	Lateral plantar nerve (S1, 2)
Plantar interossei (3)	Bases and medial shafts of 3rd, 4th, and 5th metatarsals	Medial aspects of bases of 3rd, 4th, and 5th proximal phalanxes	MP adduction and flexion of 3rd, 4th, and 5th phalanges	Cannot be palpated	Lateral plantar nerve (S1, 2)
Flexor digitorum brevis	Tuberosity of calcaneus, plantar aponeurosis	Medial and lateral aspects of 2nd, 3rd, 4th, and 5th middle phalanxes	MP and PIP flexion of 2nd, 3rd, 4th, and 5th phalanges	Cannot be palpated	Medial plantar nerve (L4, 5)

(continued)

TABLE 9.1 (continued) • **Intrinsic muscles of the foot**

Extensor digitorum brevis	Anterior and lateral calcaneus, lateral talocalcaneal ligament, inferior extensor retinaculum	Base of proximal phalanx of 1st phalange, lateral sides of extensor digitorum longus tendons of 2nd, 3rd, and 4th phalanges	Assists in MP extension of 1st phalanx and extension of middle three phalanges	Anterior to and slightly below lateral malleolus on dorsum of foot	Deep peroneal nerve (L5, S1)
Abductor digiti minimi (quinti)	Tuberosity of calcaneus, plantar aponeurosis	Lateral aspect of 5th proximal phalanx	MP abduction of 5th phalange	Cannot be palpated	Lateral plantar nerve (S1, 2)
Flexor digiti minimi (quinti) brevis	Base of 5th metatarsal, sheath of peroneus longus tendon	Lateral aspect of base of 5th proximal phalanx	MP flexion of 5th phalange	Cannot be palpated	Lateral plantar nerve (S2, 3)

Web sites

Anatomy & Physiology Tutorials:

www.gwc.maricopa.edu/class/bio201/index.htm

Radiologic Anatomy Browser:

radlinux1.usuf1.usuhs.mil/rad/iong/index.html

This site has numerous radiological views of the musculoskeletal system.

University of Arkansas Medical School Gross Anatomy for Medical Students:

anatomy.uams.edu/htmlpages/anatomyhtml/ gross.html

Dissections, anatomy tables, atlas images, links, etc.

Loyola University Medical Center: Structure of the Human Body:

www.meddean.luc.edu/lumen/MedEd/ GrossAnatomy/GA.html

An excellent site with many slides, dissections, tutorials, etc., for the study of human anatomy

Wheeless' Textbook of Orthopaedics:

www.medmedia.com/

This site has an extensive index of links to the fractures, joints, muscles, nerves, trauma, medications, medical topics, and lab tests, as well as links to orthopedic journals and other orthopedic and medical news.

Foot and Ankle Web Index:

www.footandankle.com

The foot and ankle link library located at this site is very helpful.

American College of Foot and Ankle Surgeons:

www.acfas.org

This site, sponsored by podiatric surgeons and doctors of podiatric medicine (DPM), has information on topics relating to foot health—foot and ankle deformities and injuries; care of the diabetic foot; foot and ankle disorders caused by arthritis, aging, trauma and sports injuries; and congenital deformities and disease.

The University of Texas MD Anderson Cancer Center Multimedia and Learning Resources:

rpiwww.mdacc.tmc.edu/mmlearn/anatomy.html

This site has numerous cadaveric cuts of the foot, knee, hand, arm, and elbow; an interactive ankle; and a rotating foot and ankle.

American Orthopaedic Foot and Ankle Society:

www.aofas.org

Numerous patient education brochures regarding foot and ankle problems are found here.

Premiere Medical Search Engine:

www.medsite.com

This site allows the reader to enter any medical condition and it will search the net to find relevant articles.

Virtual Hospital:

www.vh.org

Numerous slides, patient information, etc.

Dynamic Human version 2.0 CD-ROM: The Visual Guide to Anatomy & Physiology:

www.mhhe.com/biosci/ap/dynamichuman2/

Web site that accompanies this CD-ROM

Dynamic Human CD activities

1. Review anatomical landmarks as well as origins/insertions for the ankle and foot joints by clicking on **skeletal, anatomy, gross anatomy, lower limbs, tibia/fibula,** and then **foot**.
2. Review each of the muscles from this chapter by clicking on **muscular, anatomy, body regions, lower leg,** and then **foot**.

Worksheet exercise

As an aid to learning, for in-class or out-of-class assignments, or for testing, tear-out worksheets are found at the end of the text (p. 264).

Anterior and posterior skeletal worksheet (no. 1)

Draw and label on the worksheet the following muscles of the ankle and foot:
a. Tibialis anterior
b. Extensor digitorum longus
c. Peroneus longus
d. Peroneus brevis
e. Peroneus tertius
f. Soleus
g. Gastrocnemius
h. Extensor hallucis longus
i. Tibialis posterior
j. Flexor digitorum longus
k. Flexor hallucis longus

Laboratory and review exercises

1. Locate the following parts of the ankle and foot on a human skeleton and on a subject:
   a. Lateral malleolus
   b. Medial malleolus
   c. Calcaneus
   d. Navicular
   e. Three cuneiform bones
   f. Metatarsal bones
   g. Phalanges
2. How and where can the following muscles be palpated on a human subject?
   a. Tibialis anterior
   b. Extensor digitorum longus
   c. Peroneus longus
   d. Peroneus brevis
   e. Soleus
      f. Gastrocnemius
      g. Extensor hallucis longus
      h. Flexor digitorum longus
      i. Flexor hallucis longus
3. Demonstrate and palpate the following movements:
   a. Plantar flexion
   b. Dorsal flexion
   c. Inversion
   d. Eversion
   e. Flexion of the toes
   f. Extension of the toes
4. List the planes in which each of the following movements occurs. List the respective axis of rotation for each movement in each plane.
   a. Plantar flexion
   b. Dorsal flexion
   c. Inversion
   d. Eversion
   e. Flexion of the toes
   f. Extension of the toes
5. Why are "low arches" and "flat feet" not synonymous terms?
6. Discuss the value of proper footwear in various sports and activities.
7. What are orthotics and how do they function?
8. Research common foot disorders, such as flat feet, ankle injuries, and hammertoes. Report your findings in class.
9. Research the anatomical factors relating to the prevalence of inversion versus eversion ankle sprains and report your findings in class.
10. Report orally or in writing on magazine articles that rate running and walking shoes.
11. Have a laboratory partner raise up on the toes (heel raise) with the knees fully extended and then repeat with the knees flexed approximately 20 degrees. Which exercise position appears to be more difficult to maintain for an extended period of time and why? What are the implications for strengthening these muscles? For stretching these muscles?
12. Fill in the antagonistic muscle action chart by listing the muscle(s) or parts of muscles that are antagonistic in their actions to the muscles in the left column.
13. Fill in the muscle analysis chart by listing the muscles primarily involved in each joint movement.

Antagonistic muscle action chart • Ankle, transverse tarsal and subtalar joints, and toes

Agonist	Antagonist
Gastrocnemius	
Soleus	
Tibialis posterior	
Flexor digitorum longus	
Flexor hallucis longus	
Peroneus longus/ peroneus brevis	
Peroneus tertius	
Tibialis anterior	
Extensor digitorum longus	
Extensor hallucis longus	

Muscle analysis chart • Ankle, transverse tarsal and subtalar joints, and toes

Ankle	
Dorsiflexion	Plantar flexion
Transverse tarsal and subtalar joints	
Eversion	Inversion
Toes	
Flexion	Extension

References

Astrom M, Arvidson T: Alignment and joint motion in the normal foot, *Journal of Orthopaedic and Sports Physical Therapy* 22:5, November 1995.

Booher JM, Thibodeau GA: *Athletic injury assessment,* ed 4, 2000, McGraw-Hill Companies, Inc., New York.

Coughlin LP, et al: Fracture dislocation of the tarsal navicular: a case report, *American Journal of Sports Medicine* 15:614, November-December 1987.

Dearing M, Ziccardi NJ: Prevention and rehabilitation of ankle injuries, *Athletic Journal* 66:28, November 1985.

Franco AH: Pes cavus and pes planus—analysis and treatment, *Physical Therapy* 67:688, May 1987.

Gench BE, Hinson MM, Harvey PT: *Anatomical kinesiology,* Dubuque, IA, 1995, Eddie Bowers.

Grace P: Prevention and rehabilitation of shin splints, *Scholastic Coach* 57:47, March 1988.

Henderson J: Baring the soles, *Runners World* 22:14, November 1987.

Lindsay DT: *Functional human anatomy,* St. Louis, 1996, Mosby.

Luttgens K, Hamilton N: *Kinesiology: scientific basis of human motion,* ed 9, Madison, WI, 1997, Brown & Benchmark.

Robinson M: Feet first, *Coach and Athlete* 44:30, August-September 1981.

Rockar PA: The subtalar joint: anatomy and joint motion, *Journal of Orthopaedic and Sports Physical Therapy* 21:6, June 1995.

Sammarco GJ: Foot and ankle injuries in sports, *American Journal of Sports Medicine* 14:6, November-December 1986.

Seeley RR, Stephens TD, Tate P: *Anatomy & physiology,* ed 2, St. Louis, 1992, Mosby-Year Book.

Sieg KW, Adams SP: *Illustrated essentials of musculoskeletal anatomy,* ed 2, Gainesville, FL, 1985, Megabooks.

Stone RJ, Stone JA: *Atlas of the skeletal muscles,* 1990, McGraw-Hill Companies, Inc., New York.

Thibodeau GA, Patton KT: *Anatomy & physiology,* ed 9, St. Louis, 1993, Mosby.

Van De Graaff KM: *Human anatomy,* ed 4, 1995, McGraw-Hill Companies, Inc., New York.

The trunk and spinal column

10

Objectives

- To identify and differentiate the different types of vertebrae in the spinal column

- To label on a skeletal chart the types of vertebrae and their important features

- To draw and label on a skeletal chart some of the muscles of the trunk and the spinal column

- To demonstrate and palpate with a fellow student the movements of the spine and trunk and list their respective planes of motion and axes of rotation

- To palpate on a human subject some of the muscles of the trunk and spinal column

- To list and organize the muscles that produce the primary movements of the trunk and spinal column and their antagonists

The trunk and spinal column present problems in kinesiology that are not found in the study of other parts of the body. First is the complexity of the vertebral column. It consists of 24 intricate and complex articulating vertebrae. These vertebrae contain the spinal column, with its 31 pairs of spinal nerves. Unquestionably it is the most complex part of the human body other than the brain and the central nervous system.

The anterior portion of the trunk contains the abdominal muscles, which are somewhat different from other muscles in that some sections are linked by fascia and tendinous bands and thus do not attach from bone to bone. In addition, there are many small intrinsic muscles acting on the head, vertebral column, and thorax that assist in spinal stabilization or respiration, depending on their location. These muscles are generally too deep to palpate and consequently will not be given the full attention that the larger superficial muscles will receive in this chapter.

Bones

Vertebral column

The intricate and complex bony structure of the vertebral column consists of 24 articulating vertebrae and nine that are fused together (Fig. 10.1). The column is further divided into the seven cervical (neck) vertebrae, 12 thoracic (chest) vertebrae, and five lumbar (lower back) vertebrae. The sacrum (posterior pelvic girdle) and the coccyx (tail bone) consist of five and four fused vertebrae, respectively. The first two cervical vertebrae are unique in that their shapes allow for extensive rotary movements of the head to the sides, as well as forward and backward. The spine has three normal curves within its movable vertebrae. The thoracic spine curve is concave anteriorly and convex posteriorly, while the cervical and lumbar curves are concave posteriorly and convex anteriorly. The normal curves of the spine enable it to absorb blows and shocks.

The bones in each region of the spine have slightly different sizes and shapes to allow for various functions (Fig. 10.2). The vertebrae increase in size from the cervical region to the lumbar region, primarily because they have to support more weight in the lower back than in the neck. The first two cervical vertebrae are known as the atlas and axis, respectively. The vertebrae C2 through L5 have similar architecture: each has a bony block anteriorly, known as the body, a vertebral foramen centrally for the spinal cord to pass through, a transverse process projecting out laterally to each side, and a spinous process projecting posteriorly that is easily palpable.

Undesirable deviations from the normal curvatures occur due to a number of factors. Increased posterior concavity of the lumbar and cervical curves is known as *lordosis*, while increased anterior concavity of the normal thoracic curve is known as *kyphosis*. The lumbar spine may have a reduction of its normal lordotic curve, resulting in a flat-back appearance referred to as *lumbar kyphosis*. *Scoliosis* refers to lateral curvatures or sideward deviations of the spine.

Thorax

The skeletal foundation of the thorax is formed by 12 pairs of ribs (Fig. 10.3). Seven pairs are true ribs, in that they attach directly to the sternum. Five pairs are considered false ribs. Of these, three pairs attach indirectly to the sternum and two pairs are floating ribs, in that their ends are free. The manubrium, the body of the sternum, and the xiphoid process are the other bones of the thorax. All of the ribs are attached posteriorly to the thoracic vertebrae.

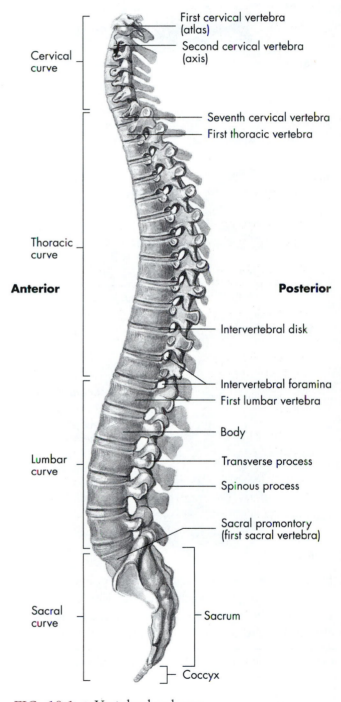

FIG. 10.1 ● Vertebral column.

From Seeley, et al: *Anatomy & physiology*, ed 3, St. Louis, 1995, Mosby.

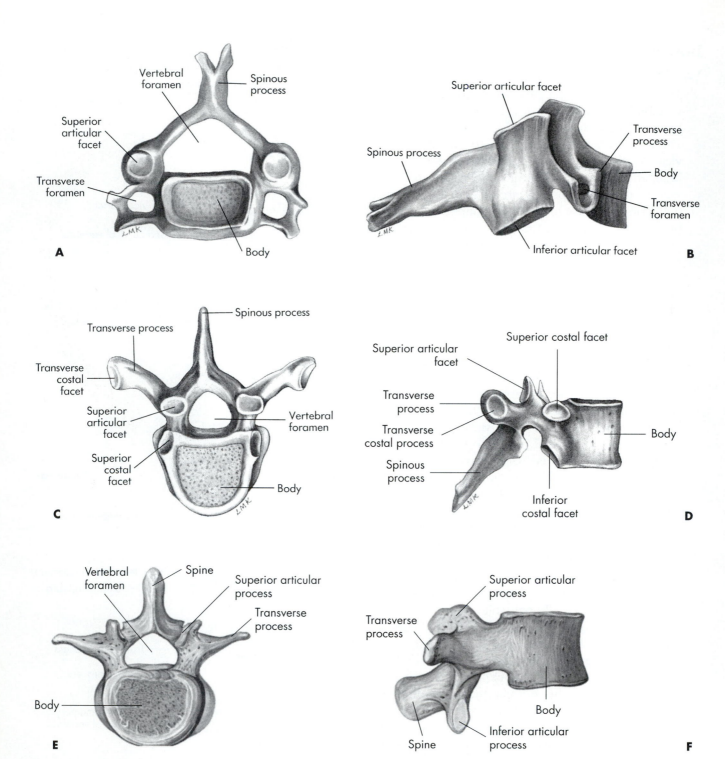

FIG. 10.2 ● Vertebral column. **A**, Typical cervical vertebra viewed from above; **B**, typical cervical vertebra viewed from the side; **C**, typical thoracic vertebra viewed from above; **D**, typical thoracic vertebra viewed from the side; **E**, third lumbar vertebra viewed from above; **F**, third lumbar vertebra viewed from the side.

From Anthony CP, Kolthoff NJ: *Textbook of anatomy and physiology,* ed 9, St. Louis, 1975, Mosby.

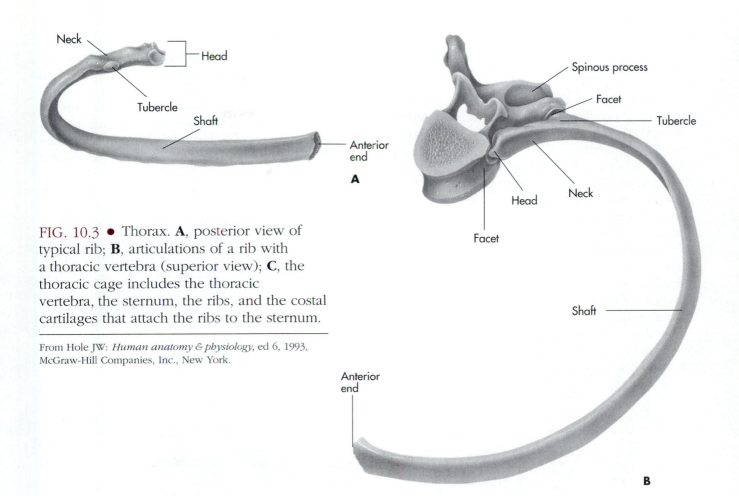

Neck

Head

Tubercle

Shaft

Anterior end

A

Spinous process

Facet

Tubercle

Neck

Head

Facet

Shaft

Anterior end

B

FIG. 10.3 ● Thorax. **A**, posterior view of typical rib; **B**, articulations of a rib with a thoracic vertebra (superior view); **C**, the thoracic cage includes the thoracic vertebra, the sternum, the ribs, and the costal cartilages that attach the ribs to the sternum.

From Hole JW: *Human anatomy & physiology*, ed 6, 1993, McGraw-Hill Companies, Inc., New York.

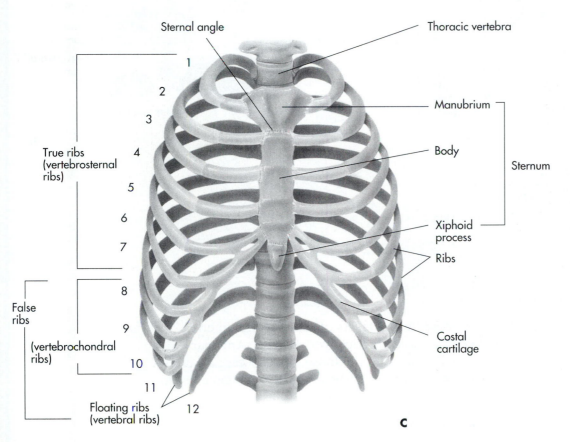

Sternal angle

Thoracic vertebra

True ribs (vertebrosternal ribs)

False ribs

(vertebrochondral ribs)

Floating ribs (vertebral ribs)

1
2
3
4
5
6
7
8
9
10
11
12

Manubrium

Body

Xiphoid process

Ribs

Costal cartilage

Sternum

C

Joints

The first joint in the axial skeleton is the atlantooccipital joint, formed by the occipital condyles of the skull sitting on the articular fossa of the first vertebra, which allows flexion and extension. The atlas (C1) in turn sits on the axis (C2) to form the atlantoaxial joint (Fig. 10.4, *A*). Except for the atlantoaxial joint, there is not a great deal of movement possible between any two vertebrae. However, the cumulative effect of combining the movement from several adjacent vertebrae allows for substantial movements within a given area. Most of the rotation within the cervical region occurs in the atlantoaxial joint, which is classified as a trochoid or pivot-type joint. The remainder of the vertebral articulations are classified as arthrodial or gliding-type joints because of their limited gliding movements.

Gliding movement occurs between the superior and inferior articular processes that form the facets joints of the vertebrae as depicted in Figs. 10.2 and 10.4, *B*. Located in between and adhering to the articular cartilage of the vertebral bodies are the intervertebral disks (Fig. 10.4, *C*). These disks are composed of an outer rim of dense fibrocartilage known as the annulus fibrosus and a central gelatinous, pulpy substance known as the nucleus pulposus. This arrangement of compressed elastic material allows compression in all directions along with torsion. With age, injury, or improper use of the spine, the intervertebral disks become less resilient, resulting in a weakened annulus fibrosus. Substantial weakening combined with compression can result in the nucleus protruding through the annulus, which is known as a herniated nucleus pulposus. Commonly referred to as a herniated or "slipped" disk, this protrusion puts pressure on the spinal nerve root, causing a variety of symptoms, including radiating pain, tingling, numbness, and weakness in the lower extremity (Fig. 10.5).

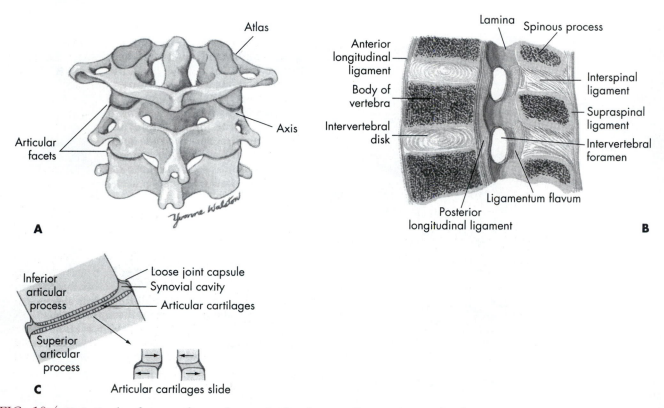

FIG. 10.4 ● Articular facets of vertebrae. **A**, the facets of superior and inferior articular processes articulate between adjacent cervical vertebrae. **B**, Articular cartilages slide back and forth on each other, and the loose articular capsule allows this motion. **C**, Ligaments limit motion between vertebrae, shown in sagittal section through three lumbar vertebrae.

From Lindsay DT: *Functional human anatomy*, St. Louis, 1996, Mosby.

A substantial number of low back problems are caused by improper use of the back over time. These improper mechanics often result in acute strains and muscle spasm of the lumbar extensors and chronic mechanical changes leading to disc herniation. Most problems occur from using the relatively small back muscles to lift objects from a lumbar spine flexed position instead of keeping the lumbar spine in a neutral position while squatting and using the larger, more powerful muscles of the lower extremity. Additionally, our lifestyles chronically place us in lumbar flexion, which over time leads to a gradual loss of lumbar lordosis. This "flat back syndrome" results in increased pressure on the lumbar disc and intermittent or chronic low back pain.

Most of the spinal column movement occurs in the cervical and lumbar regions. There is, of course, some thoracic movement, but it is slight in comparison with that of the neck and low back. In discussing movements of the head, it must be remembered that this movement occurs between the cranium and the first cervical vertebra, as well as within the other cervical vertebrae. With the understanding that these motions usually occur together, for simplification purposes this text refers to all movements of the head and neck as cervical movements. Similarly, in discussing trunk movements, lumbar motion terminology is used to describe the combined motion that occurs in both the thoracic and lumbar regions. A closer investigation of specific motion between any two vertebrae is beyond the scope of this text.

The cervical region can flex 45 degrees and extend 45 degrees. The cervical area laterally flexes 45 degrees and can rotate approximately 60 degrees. The lumbar spine, accounting for most of the trunk movement, flexes approximately 80 degrees and extends 20 to 30 degrees. Lumbar lateral flexion to each side is usually within 35 degrees, and approximately 45 degrees of rotation occurs to the left and right.

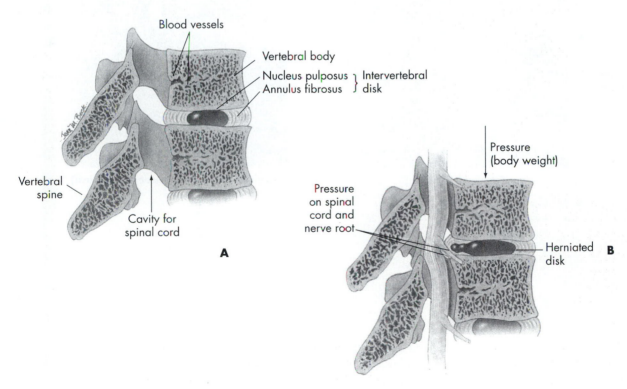

FIG. 10.5 ● Sagittal section of vertebrae showing **A,** normal disks, and **B,** herniated disks.

From Thibodeau GA, Paton KT: *Anatomy & physiology,* ed 9, St. Louis, 1993, Mosby.

Movements FIG. 10.6

Spinal movements are often preceded by the name given to the region of movement. For example, flexion of the trunk at the lumbar spine is known as lumbar flexion, and extension of the neck is often referred to as cervical extension. Additionally, as was discussed in Chapter 7, the pelvic girdle rotates as a unit due to movement occurring in the hip joints and the lumbar spine. Refer to Table 7.1.

Spinal flexion: anterior movement of the spine; in the cervical region, the head moves toward the chest; in the lumbar region, the thorax moves toward the pelvis

Spinal extension: return from flexion; posterior movement of the spine; in the cervical spine, the head moves away from the chest; in the lumbar spine, the thorax moves away from the pelvis; sometimes referred to as hyperextension

Lateral flexion (left or right): sometimes referred to as side bending; the head moves laterally toward the shoulder, and the thorax moves laterally toward the pelvis

Spinal rotation (left or right): rotary movement of the spine in the horizontal plane; the chin rotates from neutral toward the shoulder, and the thorax rotates to one side

Reduction: return movement from lateral flexion to neutral

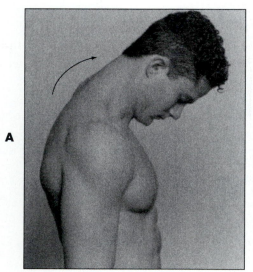

A

Cervical flexion

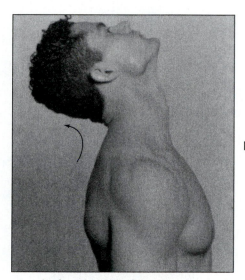

B

Cervical extension (hyperextension)

FIG. 10.6 ● Movements of the spine.

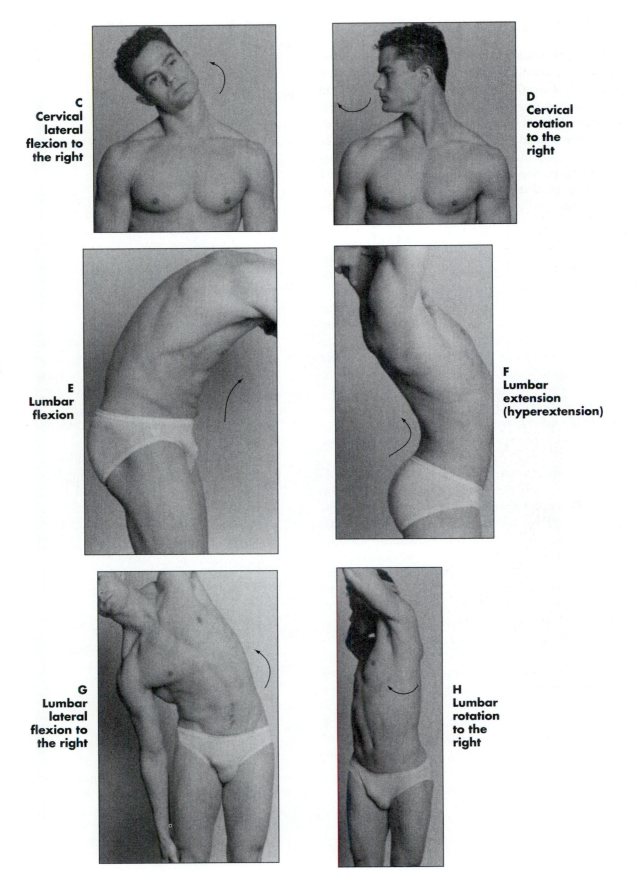

C Cervical lateral flexion to the right

D Cervical rotation to the right

E Lumbar flexion

F Lumbar extension (hyperextension)

G Lumbar lateral flexion to the right

H Lumbar rotation to the right

FIG. 10.6 continued ● Movements of the spine.

Trunk and spinal column muscles

The largest muscle in this area is the erector spinae (sacrospinalis), which extends on each side of the spinal column from the pelvic region to the cranium. It is divided into three muscles: the spinalis, the longissimus, and the iliocostalis. From the medial to the lateral side, it has attachments in the lumbar, thoracic, and cervical regions. Thus, the erector spinae group is actually made up of nine muscles. Additionally, the sternocleidomastoid and splenius muscles are large muscles involved in cervical and head movements. Large abdominal muscles involved in lumbar movements include the rectus abdominis, external oblique abdominal, internal oblique abdominal, and quadratus lumborum.

Numerous small muscles are found in the spinal column region. Many of them originate on one vertebra and insert on the next vertebra. They are important in the functioning of the spine, but knowledge of these muscles is of limited value to most people who use this text. Consequently, discussion will concentrate on the larger muscles primarily involved in trunk and spinal column movements and will only briefly address the smaller muscles.

So that the muscles of the trunk and spinal column may be better understood, they may be grouped according to both location and function. It should be noted that some muscles have multiple segments. As a result, one segment of a particular muscle may be located and perform movement in one region, while another segment of the same muscle may be located in another region to perform movements in that region. Many of the muscles of the trunk and spinal column function in moving the spine as well as in aiding respiration. All of the muscles of the thorax are primarily involved in respiration. The abdominal wall muscles are different from other muscles that have been studied. They do not go from bone to bone but attach into an aponeurosis (fascia) around the rectus abdominis area. They are the external oblique abdominal, internal oblique abdominal, and transversus abdominis.

Muscles that move the head

Anterior
  Rectus capitis anterior
  Longus capitis
Posterior
  Longissimus capitis
  Obliquus capitis superior
  Obliquus capitis inferior
  Rectus capitis posterior—major and minor
  Trapezius, superior fibers
  Splenius capitis
  Semispinalis capitis
Lateral
  Rectus capitis lateralis
  Sternocleidomastoid

Muscles of the vertebral column

Superficial
  Erector spinae (sacrospinalis)
    Spinalis—capitis, cervicis, thoracis
    Longissimus—capitis, cervicis, thoracis
    Iliocostalis—cervicis, thoracis, lumborum
  Splenius cervicis
Deep
  Longus colli—superior oblique, inferior oblique, vertical
  Interspinales—entire spinal column
  Intertransversales—entire spinal column
  Multifidus—entire spinal column
  Psoas minor
  Rotatores—entire spinal column
  Semispinalis—cervicis, thoracis

Muscles of the thorax

Diaphragm
Intercostalis—external, internal
Levator costarum
Subcostales
Scalenus—anterior, medius, posterior
Serratus posterior—superior, inferior
Transversus thoracis

Muscles of the abdominal wall

Rectus abdominis
External oblique abdominal (obliquus externus abdominis)
Internal oblique abdominal (obliquus internus abdominis)
Transverse abdominis (transversus abdominis)
Quadratus lumborum

Muscles that move the head

All muscles featured here originate on the cervical vertebrae and insert on the occipital bone of the skull, as implied by their capitis name (Figs. 10.7 and 10.8; Table 10.1). Three muscles make up the anterior vertebral muscles—the longus capitis, the rectus capitis anterior, and the rectus capitis lateralis. All are flexors of the head and upper cervical spine. The rectus capitis lateralis laterally flexes the head, in addition to assisting the rectus capitis anterior in stabilizing the atlantooccipital joint.

The rectus capitis posterior major and minor, obliquus capitis superior and inferior, and semispinalis capitis are located posteriorly. All are extensors of the head, except the obliquus capitis inferior, which rotates the atlas. The obliquus capitis superior assists the rectus capitis lateralis in lateral flexion of the head. In addition to extension, the rectus capitis posterior major is responsible for rotation of the head to the ipsilateral side. It is assisted by the semispinalis capitis, which rotates the head to the contralateral side. The splenius capitis and the sternocleidomastoid are much larger and more powerful in moving the head and cervical spine and will be covered in detail on the following pages. The remaining muscles that act on the cervical spine are addressed with the muscles of the vertebral column.

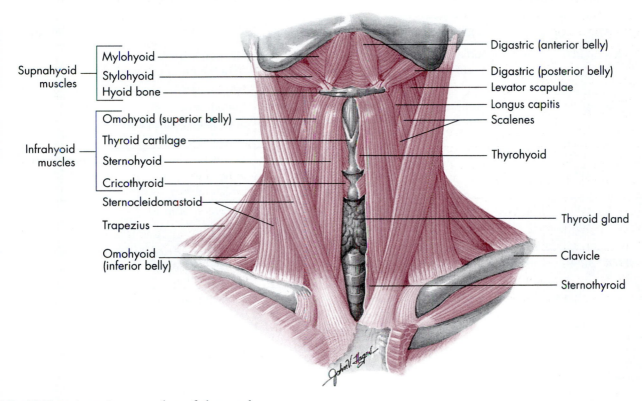

FIG. 10.7 ● Anterior muscles of the neck.

Modified from Lindsay DT: *Functional human anatomy*, St. Louis, 1996, Mosby.

FIG. 10.8 • Deep muscles of the posterior neck and upper back regions.

Modified from Van De Graaff KM: *Human anatomy*, ed 4, 1995, McGraw-Hill Companies, Inc., New York.

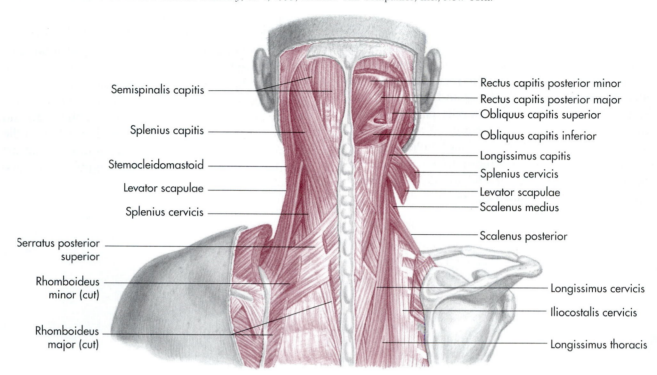

Semispinalis capitis

Splenius capitis

Sternocleidomastoid

Levator scapulae

Splenius cervicis

Serratus posterior superior

Rhomboideus minor (cut)

Rhomboideus major (cut)

Rectus capitis posterior minor

Rectus capitis posterior major

Obliquus capitis superior

Obliquus capitis inferior

Longissimus capitis

Splenius cervicis

Levator scapulae

Scalenus medius

Scalenus posterior

Longissimus cervicis

Iliocostalis cervicis

Longissimus thoracis

TABLE 10.1 • Muscles that move the head

Muscle	Origin	Insertion	Action	Innervation
Rectus capitis anterior	Anterior surface of lateral mass of atlas	Basilar part of occipital bone anterior to foramen magnum	Flexion of head and stabilization of atlanto-occipital joint	C1–3
Rectus capitis lateralis	Superior surface of transverse processes of atlas	Jugular process of occipital bone	Lateral flexion of head and stabilization of atlantooccipital joint	C1–3
Rectus capitis posterior (major)	Spinous process of axis	Lateral portion of inferior nuchal line of occipital bone	Extension and rotation of head to ipsilateral side	Posterior rami of C1
Rectus capitis posterior (minor)	Spinous process of atlas	Medial portion of inferior nuchal line of occipital bone	Extension of head	Posterior rami of C1
Longus capitis	Transverse processes of C3–6	Basilar part of occipital bone	Flexes head and cervical spine	C1–3
Obliquus capitis superior	Transverse process of atlas	Occipital bone between inferior and superior nuchal line	Extension and lateral flexion of head	Posterior rami of C1
Obliquus capitis inferior	Spinous process of axis	Transverse process of atlas	Rotation of atlas	Posterior rami of C1
Semispinalis capitis	Transverse processes of C4–T7	Occipital bone, between superior and inferior nuchal lines	Extension and contralateral rotation of head	Posterior primary divisions on spinal nerves

Sternocleidomastoid muscle   FIG. 10.9

(ster´no-kli-do-mas-toyd)

Origin

Manubrium of the sternum
Medial clavicle

Insertion

Mastoid process

Action

Both sides: flexion of the head and neck
Right side: rotation to the left and lateral flexion to
  the right

Left side: rotation to the right and lateral flexion to
  the left

Palpation

Anterolateral side of the neck, diagonally between
  the origin and insertion

Innervation

Spinal accessory nerve (Cr11, C2–3)

FIG. 10.9 ● Sternocleidomastoid muscle, anterior view.
O, Origin; *I*, insertion.

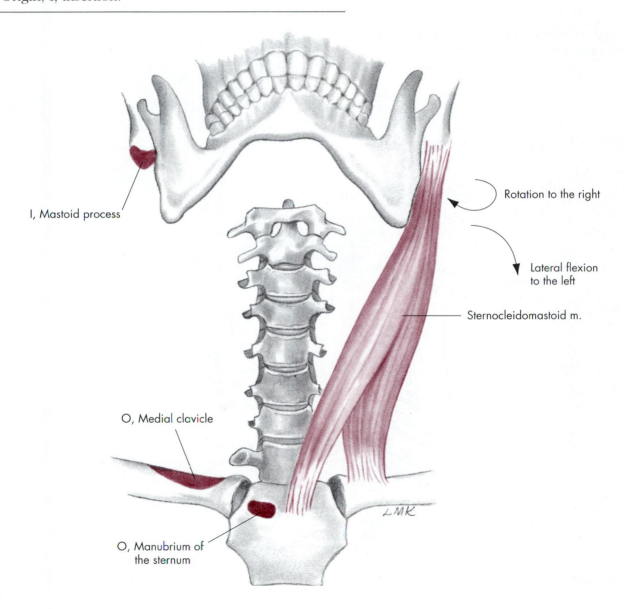

I, Mastoid process

Rotation to the right

Lateral flexion
to the left

Sternocleidomastoid m.

O, Medial clavicle

O, Manubrium of
the sternum

Application, strengthening, and flexibility

The sternocleidomastoid is primarily responsible for flexion and rotation of the head and neck. One side of this muscle may be easily visualized and palpated when rotating the head to the opposite side.

The sternocleidomastoid is easily worked for strength development by placing the hands on the forehead to apply force posteriorly while using these muscles to pull the head forward into flexion. The hand may also be used on one side of the jaw to apply rotary force in the opposite direction while the sternocleidomastoid is contracting concentrically to rotate the head in the direction of the hand.

Cervical hyperextension provides some bilateral stretching of the sternocleidomastoid. Each side may be stretched individually. The right side is stretched by moving into left lateral flexion and right cervical rotation combined with extension. The opposite movements in extension stretch the left side.

FIG. 10.9 continued ● Sternocleidomastoid muscle, lateral view. *O*, Origin; *I*, insertion.

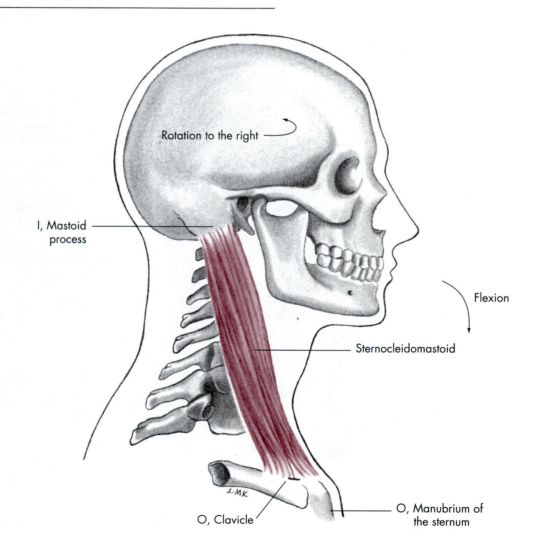

Splenius muscles (cervicis, capitis)

FIG. 10.10

(sple´ni-us) (ser´vi-sis) (kap´i-tis)

Origin

Splenius cervicis: spinous processes of the third
   through the sixth thoracic vertebrae
Splenius capitis: lower half of the ligamentum
   nuchae and the spinous processes of the seventh
   cervical and the upper three or four thoracic
   vertebrae

Insertion

Splenius cervicis: transverse processes of the first
   three cervical vertebrae
Splenius capitis: mastoid process and occipital bone

Action

Both sides: extension of the head and neck
Right side: rotation and lateral flexion to the right
Left side: rotation and lateral flexion to the left

Palpation

Cannot be palpated

Innervation

Posterior lateral branches of cervical nerves four
   through eight (C4–8)

Application, strengthening, and flexibility

Any movement of the head and neck into extension, particularly extension and rotation, would
bring the splenius muscle strongly into play, together with the erector spinae and the upper trapezius muscles. Tone in the splenius muscle tends to
hold the head and neck in proper posture position.

A good exercise for the splenius muscle is to
lace the fingers behind the head with the muscle in
flexion and then to slowly contract the posterior
head and neck muscles to move the head and neck
into full extension. This exercise may also be performed by using a towel or a partner for resistance.

The entire splenius may be stretched with maximal flexion of the head and cervical spine. The
right side can be stretched through combined
movements of left rotation, left lateral flexion, and
flexion. The same movements to the right side
apply stretch to the left side.

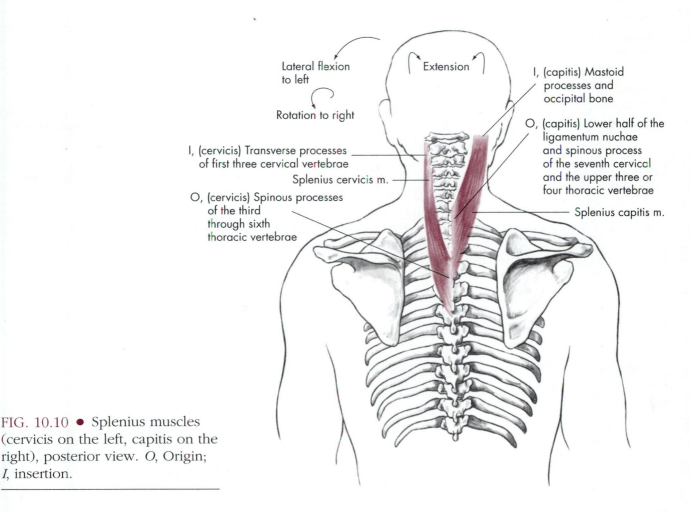

FIG. 10.10 ● Splenius muscles
(cervicis on the left, capitis on the
right), posterior view. *O,* Origin;
I, insertion.

Muscles of the vertebral column

In the cervical area, the longus colli muscles are located anteriorly and flex the cervical and upper thoracic vertebrae. Posteriorly, the erector spinae group, the transversospinalis group, the interspinal-intertransverse group, and the splenius all run vertically parallel to the spinal column (Fig. 10.11 and Table 10.2). This location enables them to extend the spine as well as assist in rotation and lateral flexion. The splenius and erector spinae group are addressed in detail elsewhere in this chapter. The transversospinalis group consists of the semispinalis, multifidus, and rotatores muscles. These muscles all originate on the transverse processes of their respective vertebrae and generally run posteriorly to attach to the spinous processes on the vertebrae just above their vertebrae of origin. All are extensors of the spine and contract to rotate their respective vertebrae to the contralateral side. The interspinal-intertransverse group lies deep to the rotatores and consists of the interspinales and the intertransversarii muscles. As a group, they laterally flex and extend but do not rotate the vertebrae. The interspinales are extensors that connect from the spinous process of one vertebra to the spinous process of the adjacent vertebra. The intertransversarii muscles flex the vertebral column laterally by connecting to the transverse processes of adjacent vertebrae.

TABLE 10.2 • Muscles of the vertebral column

Muscle	Origin	Insertion	Action	Innervation
Longus colli (superior oblique)	Transverse processes of C3–5	Anterior arch of atlas	Flexion of cervical spine	C2–7
Longus colli (inferior oblique)	Bodies of T1–3	Transverse processes of C5–6	Flexion of cervical spine	C2–7
Longus colli (vertical)	Bodies of C5–7 and T1–3	Anterior surface of bodies of C2–4	Flexion of cervical spine	C2–7
Interspinalis	Spinous process of each vertebra	Spinous process of next vertebra	Extension of spinal column	Posterior primary ramus of spinal nerves
Intertransversarii	Tubercles of transverse processes of each vertebra	Tubercles of transverse processes of next vertebra	Lateral flexion of spinal column	Anterior primary ramus of spinal nerves
Multifidus	Sacrum, iliac spine, transverse processes of lumbar, thoracic, and lower four cervical vertebra	Spinous processes of 2nd, 3rd, or 4th vertebra above origin	Extension and contralateral rotation of spinal column	Posterior primary ramus of spinal nerves
Rotatores	Transverse process of each vertebra	Base of spinous process of next vertebra above	Extension and contralateral rotation of spinal column	Posterior primary ramus of spinal nerves
Semispinalis cervicis	Transverse processes of T1–5 or 6	Spinous processes from C2–5	Extension and contralateral rotation of vertebral column	All divisions, posterior primary ramus of spinal nerves
Semispinalis thoracis	Transverse processes of T6–10	Spinous processes of C6–7 and T1–4	Extension and contralateral rotation of vertebral column	Posterior primary ramus of spinal nerves

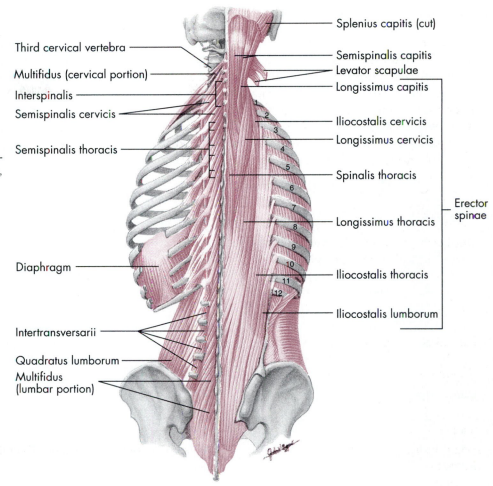

FIG. 10.11 ● Deep back muscles, posterior view. **Right**, the erector spinae group of muscles is demonstrated. **Left**, these muscles have been removed to reveal the deeper back muscles.

Modified from Seeley RR, Stephens TD, Tate P: *Anatomy & physiology,* ed 3, St. Louis, 1995, Mosby-Year Book.

Splenius capitis (cut)

Third cervical vertebra

Semispinalis capitis

Levator scapulae

Multifidus (cervical portion)

Longissimus capitis

Interspinalis

Semispinalis cervicis

Iliocostalis cervicis

Longissimus cervicis

Semispinalis thoracis

Spinalis thoracis

Longissimus thoracis

Erector spinae

Diaphragm

Iliocostalis thoracis

Iliocostalis lumborum

Intertransversarii

Quadratus lumborum

Multifidus (lumbar portion)

Muscles of Inspiration

Muscles of Expiration

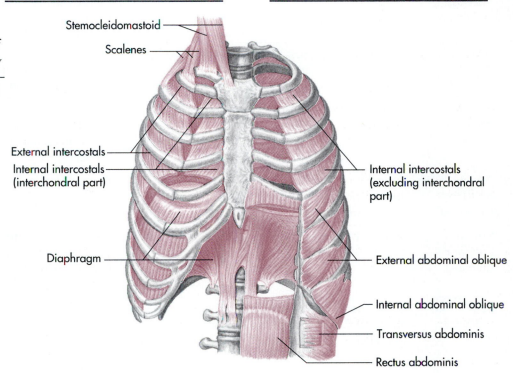

FIG. 10.12 ● Muscles of respiration, anterior view

Modified from Van De Graaff KM: *Human anatomy,* ed 4, 1995, McGraw-Hill Companies, Inc., New York.

Sternocleidomastoid

Scalenes

External intercostals

Internal intercostals (interchondral part)

Internal intercostals (excluding interchondral part)

External abdominal oblique

Diaphragm

Internal abdominal oblique

Transversus abdominis

Rectus abdominis

Muscles of the thorax

The thoracic muscles are involved almost entirely in respiration (Fig. 10.12). During quiet rest, the diaphragm is responsible for breathing movements. As it contracts and flattens, the thoracic volume is increased and air is inspired to equalize the pressure. When larger amounts of air are needed, such as during exercise, the other thoracic muscles take on a more significant role in inspiration. The scalene muscles elevate the first two ribs to increase the thoracic volume. Further expansion of the chest is accomplished by the external intercostals. Additional muscles of inspiration are the levator costarum and the serratus posterior. Forced expiration occurs with contraction of the internal intercostals, transversus thoracis, and subcostales. All of these muscles are detailed in Table 10.3.

TABLE 10.3 • **Muscles of the thorax**

Muscle	Origin	Insertion	Action	Innervation
Diaphragm	Circumference of thoracic inlet from xiphoid process, costal cartilages 6–12, and lumbar vertebrae	Central tendon of diaphragm	Depresses and draws central tendon forward in inhalation, reduces pressure in thoracic cavity, and increases pressure in abdominal cavity	Phrenic nerve (C3–5)
Internal intercostals	Longitudinal ridge on inner surface of ribs and costal cartilages	Superior border of next rib below	Elevates costal cartilages of ribs 1–4 during inhalation, depresses all ribs in exhalation	Intercostal branches of T1–11
External intercostals	Inferior border of ribs	Superior border of next rib below	Elevates ribs	Intercostal branches of T1–11
Levator costarum	Ends of transverse processes of C7, T2–12	Outer surface of angle of next rib below origin	Elevates ribs, lateral flexion of thoracic spine	Intercostal nerves
Subcostales	Inner surface of each rib near its angle	Medially on the inner surface of 2nd or 3rd rib below	Draws the ventral part of the ribs downward, decreasing the volume of the thoracic cavity	Intercostal nerves
Scalenus anterior	Transverse processes of C3–6	Inner border and upper surface of 1st rib	Elevates 1st rib, flexion, lateral flexion, and contralateral rotation of cervical spine	Ventral rami of C5–6, sometimes C4
Scalenus medius	Transverse processes of C2–7	Superior surface of 1st rib	Elevates 1st rib, flexion, lateral flexion, and contralateral rotation of cervical spine	Ventral rami of C3–8
Scalenus posterior	Transverse processes of C5–7	Outer surface of 2nd rib	Elevates 2nd rib, flexion, lateral flexion, and slight contralateral rotation of cervical spine	Ventral rami of C6–8
Serratus posterior (superior)	Ligamentum nuchae, spinous processes of C7, T1, and T2 or T3	Superior borders lateral to angles of ribs 2–5	Elevates upper ribs	Branches from anterior primary rami of T1–4
Serratus posterior (inferior)	Spinous processes of T10–12 and L1–3	Inferior borders lateral to angles of ribs 9–12	Counteracts inward pull of diaphragm by drawing last four ribs outward and downward	Branches from anterior primary rami of T9–12
Transversus thoracis	Inner surface of sternum and xiphoid process, sternal ends of costal cartilages of ribs 3–6	Inner surfaces and inferior borders of costal cartilages 3–6	Depresses ribs	Intercostal branches of T3–6

Erector spinae muscles* (sacrospinalis)

FIG. 10.13

(e-rek´tor spi´ne) (sa´kro-spi-na´lis)

Iliocostalis

(il´i-o-kos-ta´lis): lateral layer

Longissimus

(lon-jis´i-mus): middle layer

Spinalis

(spi-na´lis): medial layer

Origin

Iliocostalis: thoracolumbar aponeurosis from sacrum, posterior ribs

Longissimus: thoracolumbar aponeurosis from sacrum, lumbar and thoracic transverse processes

Spinalis: ligamentum nuchae, cervical and thoracic spinous processes.

Insertion

Iliocostalis: posterior ribs, cervical transverse processes

Longissimus: cervical and thoracic transverse processes, mastoid process

Spinalis: cervical and thoracic spinous processes, occipital bone

Action

Extension and lateral flexion of the spine

Palpation

Lower lumbar region on either side of the spine

---

*This muscle group includes the iliocostalis, the longissimus dorsi, the spinalis dorsi, and divisions of these muscles in the lumbar, thoracic, and cervical sections of the spinal column.

Innervation

Posterior branches of the spinal nerves

Application, strengthening, and flexibility

The erector spinae muscle functions best when the pelvis is posteriorly rotated. This lowers the origin of the erector spinae and makes it more effective in keeping the spine straight. As the spine is held straight, the ribs are raised, thus fixing the chest high and consequently making the abdominal muscles more effective in holding the pelvis up in front and flattening the abdominal wall.

An exercise known as the "dead lift," employing a barbell, uses the erector spinae in extending the spine. In this exercise, the subject bends over, keeping the arms and legs straight; picks up the barbell; and returns to a standing position. In performing this type of exercise, it is very important to always use correct technique to avoid back injuries. Voluntary static contraction of the erector spinae in the standing position would provide a mild exercise and improve body posture.

The erector spinae and its various divisions may be strengthened through numerous forms of back extension exercises. These are usually done in a prone or face-down position in which the spine is already in some state of flexion. The subject then uses these muscles to move part or all of the spine toward extension against gravity. A weight may be held in the hands behind the head to increase resistance.

Maximal hyperflexion of the entire spine stretches the erector spinae muscle group. Stretch may be isolated to specific segments through specific movements. Maximal flexion of the head and cervical spine stretches the capitis and cervical segments. Flexion combined with lateral flexion to one side accentuates the stretch on the contralateral side. Thoracic and lumbar flexion places the stretch primarily on the thoracis and lumborum segments.

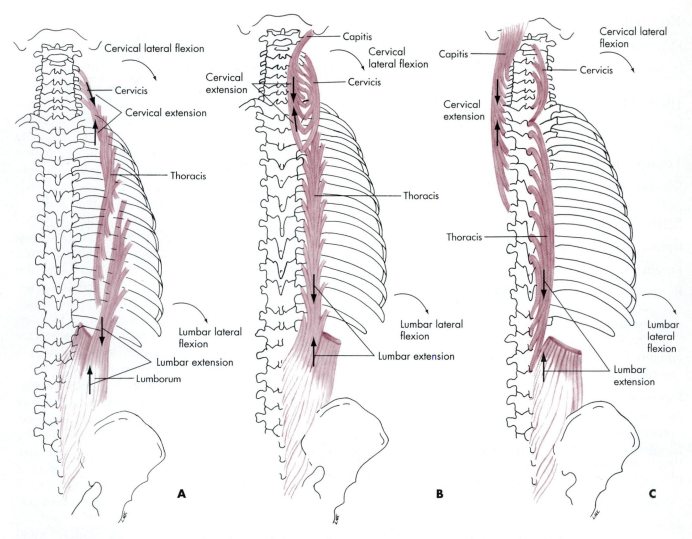

FIG. 10.13 ● Erector spinae (sacrospinalis) muscle, posterior view. **A**, Iliocostalis; **B**, longissimus; **C**, spinalis.

Muscles of the abdominal wall

FIGs. 10.14, 10.15, 10.16

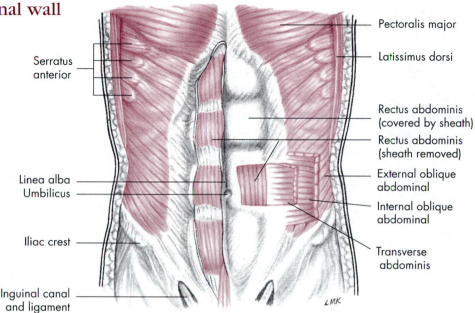

Pectoralis major

Latissimus dorsi

Serratus anterior

Rectus abdominis (covered by sheath)

Rectus abdominis (sheath removed)

External oblique abdominal

Internal oblique abdominal

Transverse abdominis

Linea alba

Umbilicus

Iliac crest

Inguinal canal and ligament

FIG. 10.14 ● Muscles of the abdomen. External oblique and rectus abdominis. The fibrous sheath around the rectus has been removed on the right side to show the muscle within.

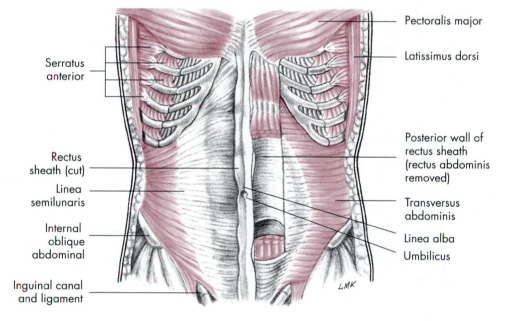

Pectoralis major

Latissimus dorsi

Serratus anterior

Rectus sheath (cut)

Linea semilunaris

Internal oblique abdominal

Inguinal canal and ligament

Posterior wall of rectus sheath (rectus abdominis removed)

Transversus abdominis

Linea alba

Umbilicus

FIG. 10.15 ● Muscles of the abdomen. The external oblique has been removed on the right to reveal the internal oblique. The external and internal obliques have been removed on the left to reveal the transversus abdominis. The rectus abdominis has been cut to reveal the posterior rectus sheath.

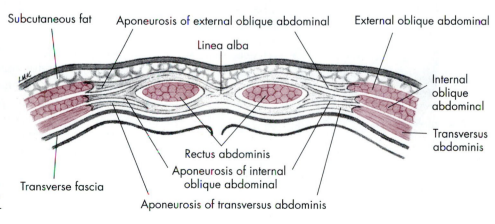

Subcutaneous fat

Aponeurosis of external oblique abdominal

Linea alba

External oblique abdominal

Internal oblique abdominal

Transversus abdominis

Rectus abdominis

Aponeurosis of internal oblique abdominal

Transverse fascia

Aponeurosis of transversus abdominis

FIG. 10.16 ● Abdominal wall. The unique arrangement of the four abdominal muscles with their fascial attachment in and around rectus abdominis muscle is shown. With no bones for attachments, these muscles can be adequately maintained through exercise.

207

Rectus abdominis muscle   FIG. 10.17

(rek´tus ab-dom´i-nis)

Origin

Crest of the pubis

Insertion

Cartilage of the fifth, sixth, and seventh ribs and the
xiphoid process

Action

Both sides: lumbar flexion
Right side: lateral flexion to the right
Left side: lateral flexion to the left

Palpation

Anteromedial surface of the abdomen, between the
rib cage and the pubic bone

Innervation

Intercostal nerves (T7–12)

Application, strengthening, and flexibility

The rectus abdominis muscle controls the tilt of
the pelvis and the consequent curvature of the
lower spine. By rotating the pelvis posteriorly, the

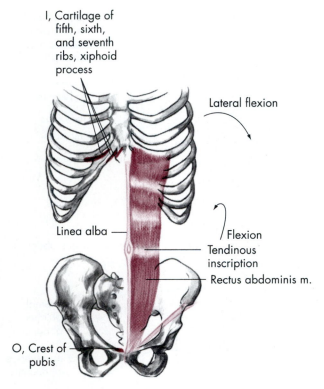

FIG. 10.17 ● Rectus abdominis muscle, anterior
view. *O,* Origin; *I,* insertion.

rectus abdominis flattens the lower back, making
the erector spinae muscle more effective as an extensor of the spine and the hip flexors (the iliopsoas muscle, particularly) more effective in raising
the legs.

In a relatively lean person with well-developed
abdominals, three distinct sets of lines or depres-
sions may be noted. Each represents an area of
tendinous connective tissue connecting or supporting the abdominal arrangement of muscles in
lieu of bony attachments. Running vertically from
the xiphoid process through the umbilicus to the
pubis is the *linea alba.* It divides each rectus abdominis and serves as its medial border. Lateral to
each rectus abdominis is the *linea semilunaris,* a
crescent, or moon-shaped, line running vertically.
This line represents the aponeurosis connecting
the lateral border of the rectus abdominis and the
medial border of the external and internal abdominal obliques. The *tendinous inscriptions* are
horizontal indentations that transect the rectus
abdominis at three or more locations, giving
the muscle its segmented appearance. Refer to
Fig. 10.16.

There are several exercises for the abdominal
muscles, such as bent-knee sit-ups, crunches, and
isometric contractions. Bent-knee sit-ups with the
arms folded across the chest are considered by
many to be a safe and efficient exercise. Crunches
are also considered to be even more effective for
isolating the work to the abdominals. Both of
these exercises shorten the iliopsoas muscle and
other hip flexors, thus reducing their ability to
generate force. Twisting to the left and right
brings the oblique muscles into more active contraction. In all of the above exercises, it is important to use proper technique, which involves
gradually moving to the up position until the lumbar spine is actively flexed maximally and then
slowly returning to the beginning position. Jerk-
ing movements using momentum should be
avoided. Movement continued beyond full lumbar
flexion only exercises the hip flexors, which is not
usually an objective. Even though all of these exercises may be helpful in strengthening the abdominals, careful analysis should occur before de-
ciding which are indicated in the presence of
various injuries and problems of the low back.

The rectus abdominis is stretched by simultaneously hyperextending both the lumbar and thoracic spine. Extending the hips assists in this
process by accentuating the anterior rotation of
the pelvis to hyperextend the lumbar spine.

External oblique abdominal muscle

FIG. 10.18

(ek-stur´nel o-bleek´ ab-dom´i-nel)

Origin

Borders of the lower eight ribs at the side of the chest, dovetailing with the serratus anterior muscle*

Insertion

Anterior half of the crest of the ilium, the inguinal ligament, the crest of the pubis, and the fascia of the rectus abdominis muscle at the lower front

Action

Both sides: lumbar flexion

Right side: lumbar lateral flexion to the right and rotation to the left

Left side: lumbar lateral flexion to the left and rotation to the right

Palpation

Lateral side of the abdomen, either left or right

---

*Sometimes the origin and insertion are reversed in anatomy books. This is the result of different interpretations of which bony structure is the more movable. The insertion is considered the most movable part of a muscle.

Innervation

Intercostal nerves (T8–12), iliohypogastric nerve (T12, L1), and ilioinguinal nerve (L1)

Application, strengthening, and flexibility

Working on each side of the abdomen, the external oblique abdominal muscles aid in rotating the trunk when working independently of each other. Working together, they aid the rectus abdominis muscle in its described action. The left external oblique abdominal muscle contracts strongly during sit-ups when the trunk rotates to the right, as in touching the left elbow to the right knee. Rotating to the left brings the right external oblique into action.

Each side of the external oblique must be stretched individually. The right side is stretched by moving into extreme left lateral flexion combined with extension or by extreme lumbar rotation to the right combined with extension. The opposite movements in extension stretch the left side.

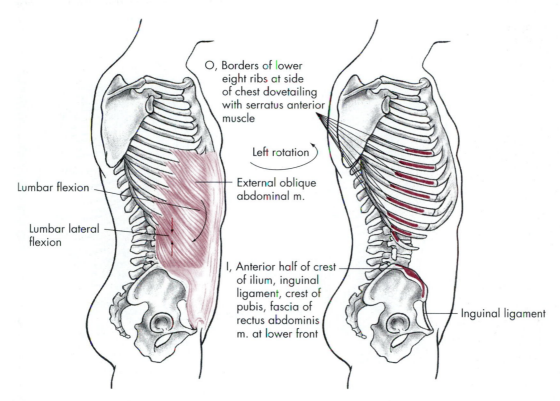

FIG. 10.18 ● External oblique abdominal muscle, lateral view. *O*, Origin; *I*, insertion.

Internal oblique abdominal muscle

FIG. 10.19

(in-ter´nel o-bleek ab-dom´i-nel)

Origin

Upper half of the inguinal ligament, anterior two-thirds of the crest of the ilium, and the lumbar fascia

Insertion

Costal cartilages of the eighth, ninth, and tenth ribs and the linea alba

Action

Both sides: lumbar flexion

Right side: lumbar lateral flexion and rotation to the right

Left side: lumbar lateral flexion and rotation to the left

Palpation

Palpated on the lateral side of the abdomen when the external oblique is relaxed

Innervation

Intercostal nerves (T8–12), iliohypogastric nerve (T12, L1), and ilioinguinal nerve (L1)

Application, strengthening, and flexibility

The internal oblique abdominal muscles run diagonally in the direction opposite to that of the external obliques. The left internal oblique rotates to the left, and the right internal oblique rotates to the right.

In touching the left elbow to the right knee in crunches, the left external oblique and the right internal oblique abdominal muscles rotate at the same time, assisting the rectus abdominis muscle in flexing the trunk to make the completion of the movement possible. In rotary movements, the internal oblique and the external oblique on opposite sides from each other always work together.

Like the external oblique, each side of the internal oblique must be stretched individually. The right side is stretched by moving into extreme left lateral flexion and extreme left lumbar rotation combined with extension. The same movements to the right combined with extension stretch the left side.

FIG. 10.19 ● Internal oblique abdominal muscle, lateral view. *O,* Origin; *I,* insertion.

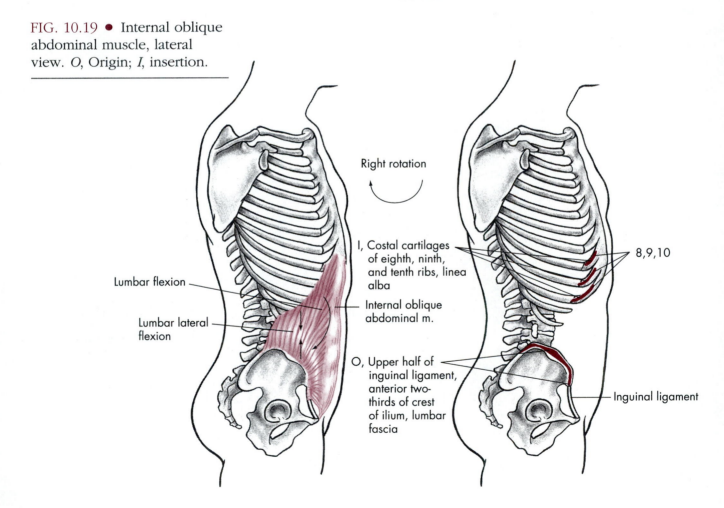

Right rotation

Lumbar flexion

Lumbar lateral flexion

I, Costal cartilages of eighth, ninth, and tenth ribs, linea alba

Internal oblique abdominal m.

O, Upper half of inguinal ligament, anterior two-thirds of crest of ilium, lumbar fascia

8,9,10

Inguinal ligament

Transversus abdominis muscle

FIG. 10.20

(trans-vurs´us ab-dom´i-nis)

Origin

Outer third of the inguinal ligament
Inner rim of the iliac crest
Inner surface of the cartilage of the lower six ribs
Lumbar fascia

Insertion

Crest of the pubis and the iliopectineal line
Abdominal aponeurosis to the linea alba

Action

Forced expiration by pulling the abdominal wall
  inward

Palpation

Cannot be palpated

Innervation

Intercostal nerves (T7–12), iliohypogastric nerve
(T12, L1), and ilioinguinal nerve (L1)

Application, strengthening, and flexibility

The transversus abdominis is the chief muscle of
forced expiration and is effective—together with
the rectus abdominis, the external oblique abdominal, and the internal oblique abdominal muscles—
in helping hold the abdomen flat. This abdominal
flattening and forced expulsion of the abdominal
contents is the only action of this muscle.

The transversus abdominis muscle is exercised
effectively by attempting to draw the abdominal
contents back toward the spine. This may be done
isometrically in the supine position or while
standing. A maximal inspiration held in the abdomen applies stretch.

FIG. 10.20 ● Transversus abdominis muscle, lateral view. *O*, Origin; *I*, insertion.

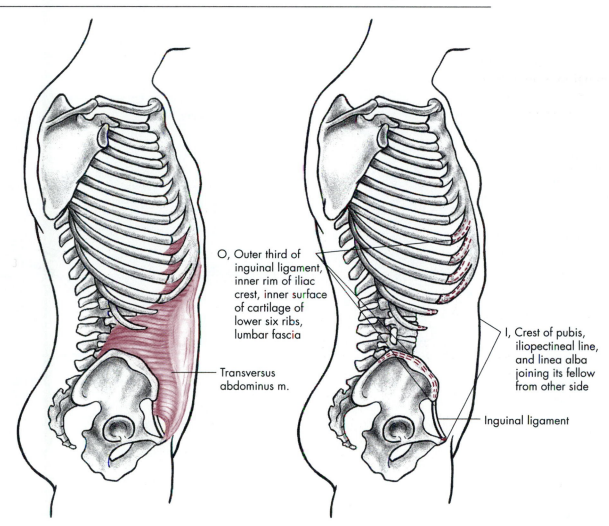

O, Outer third of inguinal ligament, inner rim of iliac crest, inner surface of cartilage of lower six ribs, lumbar fascia

Transversus abdominus m.

I, Crest of pubis, iliopectineal line, and linea alba joining its fellow from other side

Inguinal ligament

Quadratus lumborum muscle FIG. 10.21

(kwad-ra´tus lum-bo´rum)

Origin

Posterior inner lip of the iliac crest

Insertion

Approximately one-half the length of the lower
  border of the twelfth rib and the transverse process
  of the upper four lumbar vertebrae

Action

Lateral flexion to the side on which it is located
Stabilizes the pelvis and lumbar spine

Palpation

For all practical purposes, impossible to palpate,
  except on an extremely thin individual

Innervation

Branches of T12, L1 nerves

Application, strengthening, and flexibility

The quadratus lumborum is important in lumbar
lateral flexion and in elevating the pelvis on the
same side in the standing position. Trunk rotation
and lateral flexion movements against resistance
are good exercises for development of this muscle. The position of the body relative to gravity
may be changed to increase resistance on this and
other trunk and abdominal muscles. Left lumbar
lateral flexion stretches the right quadratus lumborum and vice versa.

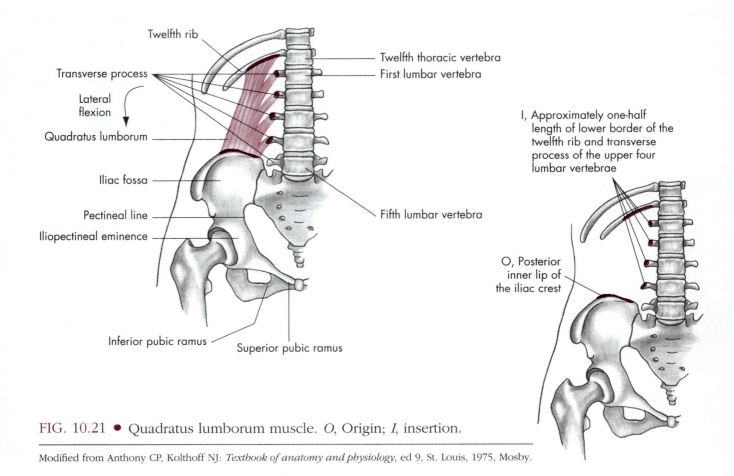

FIG. 10.21 ● Quadratus lumborum muscle. *O,* Origin; *I,* insertion.

Modified from Anthony CP, Kolthoff NJ: *Textbook of anatomy and physiology,* ed 9, St. Louis, 1975, Mosby.

Web sites

Anatomy & Physiology Tutorials:

www.gwc.maricopa.edu/class/bio201/index.htm

Radiologic Anatomy Browser:

radlinux1.usuf1.usuhs.mil/rad/iong/index.html

This site has numerous radiological views of the musculoskeletal system.

Loyola University Medical Center: Structure of the Human Body:

www.meddean.luc.edu/lumen/MedEd/ GrossAnatomy/GA.html

An excellent site with many slides, dissections, tutorials, etc., for the study of human anatomy

University of Arkansas Medical School Gross Anatomy for Medical Students:

anatomy.uams.edu/htmlpages/anatomyhtml/ gross.html

Dissections, anatomy tables, atlas images, links, etc.

Wheeless' Textbook of Orthopaedics:

www.medmedia.com/

This site has an extensive index of links to the fractures, joints, muscles, nerves, trauma, medications, medical topics, and lab tests, as well as links to orthopedic journals and other orthopedic and medical news.

Premiere Medical Search Engine:

www.medsite.com

This site allows the reader to enter any medical condition and it will search the net to find relevant articles.

Virtual Hospital:

www.vh.org

Numerous slides, patient information, etc.

Dynamic Human version 2.0 CD-ROM: The Visual Guide to Anatomy & Physiology:

www.mhhe.com/biosci/ap/dynamichuman2/

Web site that accompanies this CD-ROM

Dynamic Human CD activities

1. Review anatomical landmarks as well as origins/insertions for the trunk and spinal column by clicking on **skeletal, anatomy, gross anatomy, vertrebral column, thoracic cage**, and then **pelvic girdle**.
2. Review each of the muscles from this chapter by clicking on **muscular, anatomy, body regions, head and neck**, and then **abdomen and back**.

3. Think about how prolonged and repetitive flexion may affect a herniated disc. What about extension? Click on **skeletal, clinical concepts**, and then **herniated disc**.

Worksheet exercises

As an aid to learning, for in-class or out-of-class assignments, or for testing, tear-out worksheets are found at the end of the text (pp. 265 and 266).

Anterior skeletal worksheet (no. 1)

Draw and label the following muscles on the skeletal chart:
a. Rectus abdominis
b. External oblique abdominal
c. Internal oblique abdominal

Posterior skeletal worksheet (no. 2)

Draw and label the following muscles on the skeletal chart:
a. Erector spinae
b. Quadratus lumborum
c. Splenius—cervicis and capitis

Laboratory and review exercises

1. Locate the following parts of the spine on a human skeleton and on a human subject:
   a. Cervical vertebrae
   b. Thoracic vertebrae
   c. Lumbar vertebrae
   d. Spinous processes
   e. Transverse processes
   f. Sacrum
   g. Manubrium
   h. Xiphoid process
   i. Sternum
   j. Rib cage (various ribs)
2. How and where can the following muscles be palpated on a human subject?
   a. Rectus abdominis
   b. External oblique abdominal
   c. Internal oblique abdominal
   d. Erector spinae
   e. Sternocleidomastoid
3. List the planes in which each of the following movements occurs. List the respective axis of rotation for each movement in each plane.
   a. Cervical flexion
   b. Cervical extension
   c. Cervical rotation
   d. Cervical lateral flexion
   e. Lumbar flexion
   f. Lumbar extension

g. Lumbar rotation

h. Lumbar lateral flexion

4. Contrast crunches with bent-knee sit-ups and with straight-leg sit-ups. Does having a partner to hold the feet make a difference in the ability to do the bent-knee and straight-leg sit-ups? If so, why?

5. Have a laboratory partner stand and assume a position exhibiting good posture. What motions in each region of the spine does gravity attempt to produce? Which muscles are responsible for counteracting these motions against the pull of gravity?

6. Compare and contrast the spinal curves of a laboratory partner sitting erect versus one sitting slouched in a chair. Which muscles are responsible for maintaining good sitting posture?

7. Which exercise is better for the development of the abdominal muscles—leg-lifts or sit-ups? Analyze each one in regard to the activity of the abdominal muscles. Defend your answer.

8. Why is good abdominal muscular development so important? Why is this area so frequently neglected?

9. Why are weak abdominal muscles frequently blamed for lower back pain?

10. Prepare an oral or a written report on abdominal or back injuries found in the literature.

11. Fill in the muscle analysis chart by listing the muscles primarily involved in each movement.

12. Fill in the antagonistic muscle action chart by listing the muscle(s) or parts of muscles that are antagonistic in their actions to the muscles in the left column.

Muscle analysis chart • Cervical and lumbar spine

Cervical spine	
Flexion	Extension
Lateral flexion right	Rotation right
Lateral flexion left	Rotation left
Lumbar Spine	
Flexion	Extension
Lateral flexion right	Rotation right
Lateral flexion left	Rotation left

Antagonistic muscle action chart • Cervical and lumbar spine

Agonist	Antagonist
Splenius capitis	
Splenius cervicis	
Sternocleidomastoid	
Erector spinae	
Rectus abdominis	
External oblique abdominal	
Internal oblique abdominal	
Quadratus lumborum	

References

Clarkson HM, Gilewich GB: *Musculoskeletal assessment: joint range of motion and manual muscle strength,* Baltimore, 1989, Williams & Wilkins.

Day AL: Observation on the treatment of lumbar disc disease in college football players, *American Journal of Sports Medicine,* 15:275, January-February 1987.

Gench BE, Hinson MM, Harvey PT: *Anatomical kinesiology,* Dubuque, IA, 1995, Eddie Bowers.

Hislop HJ, Montgomery J: *Daniels and Worthingham's muscle testing: techniques of manual examination,* ed 6, Philadelphia, 1995, Saunders.

Holden DL, Jackson DW: Stress fractures of ribs in female rowers, *American Journal of Sports Medicine,* 13:277, July-August 1987.

Lindsay DT: *Functional human anatomy,* St. Louis, 1996, Mosby.

Luttgens K, Hamilton N: *Kinesiology: scientific basis of human motion,* ed 9, Madison, WI, 1997, Brown & Benchmark.

Martens MA, et al: Adductor tendonitis and muscular abdominis tendopathy, *American Journal of Sports Medicine,* 15:353, July-August 1987.

Marymont JV: Exercise-related stress reaction of the sacroiliac joint, an unusual cause of low back pain in athletes, *American Journal of Sports Medicine,* 14:320, July-August 1986.

Perry JF, Rohe DA, Garcia AO: *The kinesiology workbook,* Philadelphia, 1992, Davis.

Rasch PJ: *Kinesiology and applied anatomy,* ed 7, Philadelphia, 1989, Lea & Febiger.

Seeley RR, Stephens TD, Tate P: *Anatomy & physiology,* ed 2, St. Louis, 1992, Mosby-Year Book.

Sieg KW, Adams SP: *Illustrated essentials of musculoskeletal anatomy,* ed 2, Gainesville, FL, 1985, Megabooks.

Stone RJ, Stone JA: *Atlas of the skeletal muscles,* 1990, McGraw-Hill Companies, Inc., New York.

Thibodeau GA, Patton KT: *Anatomy & physiology,* ed 9, St. Louis, 1993, Mosby.

Van De Graaff KM: *Human anatomy,* ed 4, 1995, McGraw-Hill Companies, Inc., New York.

Muscular analysis of trunk and lower extremity exercises

Objectives

- To understand various conditioning principles and how to apply them to strengthening major muscle groups

- To apply the concept of the kinetic chain to the lower extremity

- To analyze an exercise for the types of contractions occurring in specific muscles to create major joint movements

- To discuss the various types of exercise equipment used in developing muscular strength and endurance

- To learn to analyze and prescribe exercises to strengthen all major muscle groups

Chapter 6 presented an introduction to the analysis of exercise and activities. That chapter included only the analysis of the muscles previously studied in the upper extremity region.

Since that chapter, all the other joints and large muscle groups of the human body have been considered. The exercises and activities found in this chapter concentrate more on the muscles in the trunk and lower extremity.

Strength, endurance, and flexibility of the muscles of the lower extremity, trunk, and abdominal sections are also very important in skillful physical performance and body maintenance.

The type of contraction is determined by whether the muscle is lengthening or shortening during the movement. However, muscles may shorten or lengthen in the absence of a contraction through passive movement caused by other contracting muscles, momentum, gravity, or external forces such as manual assistance and exercise machines.

A concentric contraction is a shortening contraction of the muscles against gravity or resistance, whereas an eccentric contraction is a contraction in which the muscle lengthens under tension to control the joints moving with gravity or resistance.

Contraction against gravity is also quite evident in the lower extremities.

The quadriceps muscle group contracts eccentrically when the body slowly lowers in a weight-bearing movement through lower extremity action. The quadriceps functions as a decelerator to knee joint flexion in weight-bearing movements by contracting eccentrically to prevent too rapid of a downward movement. One can easily demonstrate this fact by palpating this muscle group when slowly moving from a standing position to a half-squat. Almost as much work is done in this type of contraction as in concentric contractions.

In this example involving the quadriceps, the slow descent is eccentric, and the ascent from the squatted position is concentric. If the descent were under no muscular control, it would be at the same speed as gravity and the muscle lengthening would be passive. That is, the movement and change in length of the muscle would be caused by gravity and not by active muscular contractions.

In recent years, more and more muscle educators on all levels have been emphasizing the development of muscle groups through resistance training and circuit-training activities.

Athletes and nonathletes, both male and female, need to have overall muscular development.

Sport participation does not ensure sufficient development of muscle groups. Also, more and more emphasis has been placed on mechanical kinesiology in physical education and athletic skill teaching. This is desirable and can help bring about more skillful performance. However, one must remember that mechanical principles will be of little or no value to performers without adequate strength and endurance of their muscular system, which is developed through planned exercises and activities. In the fitness and health revolution of recent years, a much greater emphasis has been placed on exercises and activities that improve the physical fitness, strength, endurance, and flexibility of participants. This chapter will continue the practice of analyzing the muscles through simple exercises that began in Chapter 6. When these techniques are mastered, the individual is ready to analyze and prescribe exercises and activities for muscular strength and endurance needed in sport activities and for healthful living.

Conditioning considerations

Overload principle

A basic physiological principle of exercise is the overload principle. It states that, within appropriate parameters, a muscle or muscle group increases in strength in direct proportion to the overload placed on it. While it is beyond the scope of this text to fully explain specific applications of the overload principle for each component of physical fitness, some general concepts will be mentioned. To improve the strength and functioning of major muscles, educators need to apply this principle to every large muscle group in the body, progressively throughout each year, at all age levels. In actual practice, the amount of overload applied varies significantly based on several factors. For example, an untrained person beginning a strength training program will usually make significant gains in the amount of weight he or she is able to lift in the first few weeks of the exercise program. Most of this increased ability is due to a refinement of neuromuscular function, rather than an actual increase in muscle tissue strength. Similarly, a well-trained person will see relatively minor improvements in the amount of weight that can be lifted over a much longer period of time. Therefore, the amount and rate of progressive overload are extremely variable and must be adjusted to match the specific needs of the individual's exercise objectives. Overload is not always progressively increased. In certain periods of conditioning, the overload should actually be prescriptively reduced or increased to improve the total results of the entire program. Overload may be modified by changing any one or a combination of three exercise variables—*frequency, intensity,* and *duration*. Increasing the speed of doing the exercise, the number of repetitions, the weight, and more bouts of exercise are all ways to modify these variables and apply this principle.

SAID principle

The SAID (**S**pecific **A**daptations to **I**mposed **D**emands) principle should be considered in all aspects of physiological conditioning and training. This principle, which states that the body will gradually, over time, adapt very specifically to the various stresses and overloads to which it is subjected, is applicable in every form of muscle training, as well as to the other systems of the body. For example, if an individual were to undergo several weeks of strength-training exercises for a particular joint through a limited range of motion, the specific muscles involved in performing the strengthening exercises would improve primarily in the ability to move against increased resistance through the specific range of motion used. There would be, in most cases, minimal strength gains beyond the range of motion used in the training. Additionally, other components of physical fitness—such as flexibility, cardiorespiratory endurance, and muscular endurance—would be enhanced minimally, if at all. In other words, to achieve specific benefits, exercise programs must be specifically designed for the adaptation desired.

It should be recognized that this adaptation may be positive or negative, depending on whether or not the correct techniques are used and stressed in the design and administration of the conditioning program. Inappropriate or excessive demands placed on the body in too short of a time span can result in injury. If the demands are too minimal or administered too infrequently over too long a time period, less than desired improvement will occur. Conditioning programs and the exercises included in them should be analyzed to determine if they are using the specific muscles for which they were intended in the correct manner.

Specificity

Specificity of exercise strongly relates to the discussion of the SAID principle. The components of physical fitness, such as muscular strength, muscular endurance, and flexibility, are not general body characteristics. They are specific to each body area and muscle group. Therefore, the specific needs of the individual must be addressed when designing an exercise program. Quite often, it will be necessary to analyze an individual's exercise and skill technique to design an exercise program to meet his or her specific needs. Potential exercises to be used in the conditioning program must be analyzed to determine their appropriateness for the individual's specific needs. The goals of the exercise program should be determined regarding specific areas of the body, preferred time to physically peak, and physical fitness needs such as strength,

muscular endurance, flexibility, cardiorespiratory endurance, and body composition. After establishing goals, a regimen incorporating the overload variables of frequency, intensity, and duration may be prescribed to include the entire body or specific areas in a manner to address the improvement of the preferred physical fitness components. Regular observation and follow-up exercise analysis is necessary to ensure proper adherence to correct technique.

Muscular development

For years it was thought that a person developed adequate muscular strength, endurance, and flexibility through participation in sport activities. Now the philosophy is that one needs to develop muscular strength, endurance, and flexibility in order to be able to participate safely and effectively in sport activities.

Adequate muscular strength, endurance, and flexibility of the entire body from head to toe should be developed through correct use of the appropriate exercise principles. Individuals responsible for this development need to prescribe exercises that will meet these objectives.

In schools this development should start at an early age and continue throughout the school years. Results of fitness tests reveal the need for considerable improvement in this area. The chin-up (pull-up) test had to be modified because more than 50% of children could not do one chin-up. Sit-ups, the standing long jump, the mile run, and other tests all indicated fitness deficiencies in the children of the United States. Adequate muscular strength and endurance are important in the adult years for the activities of daily living, as well as job-related requirements and recreational needs. Many back problems and other physical ailments could be avoided through proper maintenance of the musculoskeletal system.

The sampling of exercises in this chapter will assist individuals in analyzing and prescribing exercises for overall muscular development of the trunk and lower extremities in all age groups. All muscles listed in the analysis are contracting concentrically unless specifically noted to be contracting isometrically or eccentrically. Chapter 6 includes several exercises for the upper extremities.

Sit-up—bent knee   FIG. 11.1

Description

The participant lies on the back, fingertips beside the head, with the knees flexed approximately 90 degrees and the feet about hip-width apart. The hips and knees are flexed in this manner to reduce the length of the hip flexors, thereby reducing their contribution to the sit-up. This positioning will allow more emphasis on the abdominals, as compared with a straight leg sit-up.

The participant curls up to a sitting position, twists the trunk to the left, touches the right elbow to the left knee, and then returns to the starting position. On the next repetition, the participant should twist to the right instead of the left for balanced muscular development.

Analysis

This exercise is divided into four movements for analysis: (1) curling movement to sitting-up position, (2) twisting movement to left, (3) return movement to sitting-up position, and (4) return movement to starting position.

Curling movement to sitting-up position
  Trunk and cervical spine
    Flexion
      Trunk and cervical spine flexors
        Rectus abdominis
        External oblique abdominal
        Internal oblique abdominal
        Sternocleidomastoid

Hip
  Flexion
    Hip flexors
      Iliopsoas
      Rectus femoris
      Pectineus
Twisting movement to left
  Trunk and cervical spine
    Left lumbar rotation
      Left rectus abdominis
      Right external oblique abdominal
      Left internal oblique abdominal
      Left erector spinae
Return movement to sitting-up position
  Trunk
    Right lumbar rotation
      Right rectus abdominis
      Left external oblique abdominal
      Right internal oblique abdominal
      Right erector spinae
Return movement to starting position
  Trunk
    Extension
      Trunk and cervical spine flexors
      (eccentric contraction)
  Hip
    Extension
      Hip flexors (eccentric contraction)

NOTE: Slight movement of the shoulder joint and girdle is not being analyzed.

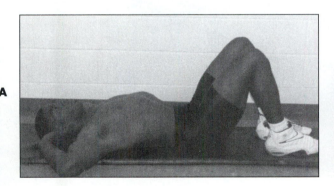
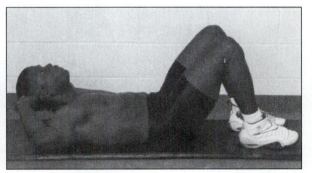

FIG. 11.1 ● Sit-up, bent knee. **A**, Beginning relaxed position; **B**, up position.

Prone arch   FIG. 11.2

Description

The participant lies in a prone position, face down, with the arms in an adducted and relaxed position lying beside the body. The head, upper trunk, and thighs are raised from the floor. The knees are kept in full extension. Then the participant returns to the starting position.

Analysis

This exercise is separated into two movements for analysis: (1) movement to raise the head, trunk, and thighs and (2) return movement to starting position.

Movement to raise the head, trunk, and thighs
   Trunk and cervical spine
      Extension
         Trunk and cervical spine extensors
            Erector spinae
            Splenius
            Quadratus lumborum

Hip
      Extension
         Hip extensors
            Gluteus maximus
            Semitendinosus
            Semimembranosus
            Biceps femoris
Return movement to starting position
   Trunk and cervical spine
      Flexion (return to neutral flat position)
         Trunk and cervical spine extensors
          (eccentric contraction)
Hip
      Flexion (return to neutral flat position)
         Hip extensors (eccentric contraction)

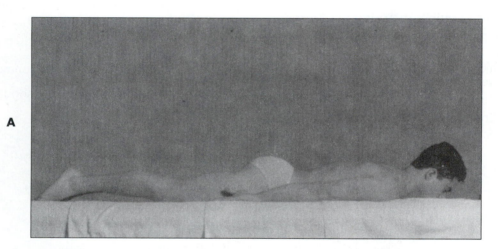

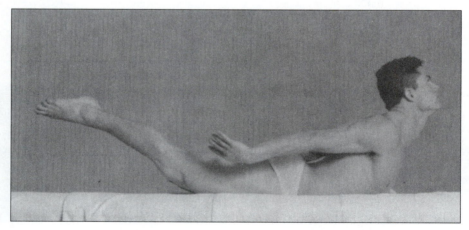

FIG. 11.2 ● Prone arch. **A,** Beginning relaxed position; **B,** fully arched position.

Free weight–training exercises*

Exercise through the use of weights has become increasingly important as a means of developing and maintaining muscular strength in young people and adults. When this type of exercise is undertaken, a thorough knowledge of the muscles being used is essential so that one group of muscles is not overdeveloped and another underdeveloped.

Most schools have free weights, barbells, and dumbbells available for use by students. Some physical education teachers and coaches recommend that junior and senior high school students have their own barbell set for use at home.

An analysis of several selected weight-training exercises will introduce the muscular analysis of these activities. In these exercises, the only equipment needed is a barbell with weights.

---

*These are only sample weight-training exercises. Students are encouraged to continue the study of muscular analysis of the many other weight-training exercises and activities.

Squat   FIG. 11.3

Description

The participant places a barbell on the shoulders behind the neck and grasps it with the palms-forward position of the hands. The participant squats down, flexing at the hips while keeping the spine in normal alignment, until the thighs are parallel to the floor. The participant then returns to the starting position. This exercise is commonly performed improperly by allowing the knees to move forward beyond the plane of the feet, which greatly increases the risk of injury. Care should be taken to ensure that the shins remain as vertical as possible during this exercise.

The feet should be parallel, with slight external rotation of the lower extremity. The knees should point over the ankles and feet without going in front, between, or outside of the vertical plane of feet.

Analysis

This exercise is separated into two movements for analysis: (1) movement to the knee-bend position and (2) return movement to starting position. NOTE: It is assumed that no movement will take place in the shoulder joint, shoulder girdle, wrists, hands, and back.

Movement to knee-bend position
  Hip
    Flexion
      Hip extensors (eccentric contraction)
        Gluteus maximus
        Semimembranosus
        Semitendinosus
        Biceps femoris
  Knee
    Flexion
      Knee extensors (eccentric contraction)
        Rectus femoris
        Vastus lateralis
        Vastus medialis
        Vastus intermedius

Foot and ankle
  Dorsal flexion
    Plantar flexors (eccentric contraction)
      Gastrocnemius
      Soleus
Return movement to starting position
  Hip
    Extension
      Hip extensors
  Knee
    Extension
      Knee extensors
  Foot and ankle
    Plantar flexion
      Plantar flexors

FIG. 11.3 • Squat. **A**, Starting position; **B**, squatted position.

Dead lift   FIG. 11.4

Description

The participant begins in hip flexed position, keeping the arms, legs, and back straight, and grasps the barbell on the floor. Then a movement to the standing position is made by extending the hips. This exercise, when done improperly by allowing lumbar flexion, may contribute to low back problems. It is essential that the lumbar extensors be used more as isometric stabilizers of the low back while the hip extensors perform the majority of the lift in this exercise.

Analysis

This exercise is divided into two movements for analysis: (1) movement to standing position and (2) return movement to place the barbell on the floor.

Movement to standing position
    Wrist and hand
        Flexion
            Wrist and hand flexors (isometric contraction)
                Flexor carpi radialis
                Flexor carpi ulnaris
                Palmaris longus
                Flexor digitorum profundus
                Flexor digitorum superficialis
                Flexor pollicis longus
    Trunk
        Extension
            Trunk extensors (isometric contraction)
                Erector spinae (sacrospinalis)
                Quadratus lumborum

Hip
    Extension
        Hip extensors
            Gluteus maximus
            Semimembranosus
            Semitendinosus
            Biceps femoris
Knee
    Extension
        Knee extensors
            Rectus femoris
            Vastus lateralis
            Vastus medialis
            Vastus intermedius
Return movement to place the barbell on floor
    Wrist and hand
        Flexion
            Wrist and hand flexors (isometric contraction)
    Trunk
        Extension
            Trunk extensors (isometric contraction)
    Hip
        Flexion
            Hip extensors (eccentric contraction)
    Knee
        Flexion
            Knee extensors (eccentric contraction)

NOTE: Slight movement of the shoulder joint and girdle is not being analyzed.

A   **B**

FIG. 11.4 ● Dead lift. **A**, Beginning hip flexed position; **B**, ending hip extendend position.

Isometric exercises

An exercise technique called "isometrics" is a type of muscular activity in which there is contraction of muscle groups with little or no muscle shortening. Although not as productive in terms of overall strength gains as isotonics, isometrics are an effective way to build and maintain muscular strength in a limited range of motion.

A few selected isometric exercises are analyzed muscularly to show how they are designed to develop specific muscle groups. Although there are varying approaches to isometrics, most authorities agree that isometric contractions should be held approximately 7 to 10 seconds for a training effect.

Abdominal contraction FIG. 11.5

Description

The participant contracts the muscles in the anterior abdominal region as strongly as possible, with no movement of the trunk or hips. This exercise can be performed in sitting, standing, or supine positions. The longer the contraction in seconds, the more valuable the exercise will be, to a degree.

Analysis

Abdomen
    Contraction
        Rectus abdominis
        External oblique abdominal
        Internal oblique abdominal
        Transversus abdominis

FIG. 11.5 ● Abdominal contraction. **A**, Beginning position; **B**, contracted position.

Leg lifter   FIG. 11.6

Description

The participant sits on a bench or chair with the knees slightly bent and with the left leg over the right. An attempt to raise the right leg while resisting it with the left leg is conducted.

Analysis*

Right leg—attempt upward movement
    Foot and ankle
        Dorsal flexion
            Tibialis anterior
            Extensor hallucis longus
            Extensor digitorum longus
            Peroneus tertius
    Knee
        Extension
            Quadriceps
                Rectus femoris
                Vastus lateralis
                Vastus medialis
                Vastus intermedius

*When the legs are alternated, the muscles used will be the same muscles but in the other leg.

Hip
    Flexion
        Iliopsoas
        Rectus femoris
        Pectineus
        Sartorius
        Tensor fasciae latae
Left leg—resisting upward movement
    Foot and ankle
        Plantar flexion
            Gastrocnemius
            Soleus
    Knee
        Flexion
            Hamstrings
                Biceps femoris
                Semitendinosus
                Semimembranosus
    Hip
        Extension
            Gluteus maximus
            Biceps femoris
            Semitendinosus
            Semimembranosus

FIG. 11.6 ● Leg lifter. **A**, Beginning relaxed position; **B**, up position.

Total Gym exercise machine

The Total Gym exercise machine* or similar machines by other manufacturers (Fig. 11.7) are used by professional, college, and many high school athletes, as well as thousands of home users. Health clubs, YMCAs, YWCAs, fitness centers, and body gyms have these machines for use by their members. Many similar types of machines are available; these include Cybex Eagle, Universal, Nautilus, Nordic Flex Ultralift, Paramount, Olympus, and others. Many of these machines rely on various pulleys, cams, and positions to vary the resistance, as opposed to free weights, which were discussed earlier in this chapter.

All exercise machines come with a list of recommended exercises that can be done by the user. The advantages of these machines are that they may be used individually to perform a wide variety of exercises for many body areas, regardless of the strength level. See fig. 11.7 for a few of the exercises for the Total Gym.

*Efi medical systems, San Diego, CA.

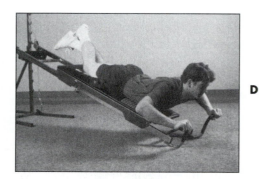

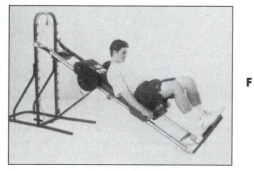

FIG. 11.7 ● Total Gym exercise machine. **A**, biceps curl; **B**, single-leg hip extension; **C**, pull-up; **D**, incline press; **E**, alternating shoulder punch; **F**, squat; **G**, dip.

Leg press   FIG. 11.8

Description

The participant sits and presses until the knees are straight, then returns to the starting position.

Analysis

Leg press can be divided into two movements for analysis: (1) movement to straight-leg position and (2) return to standing position.

Movement to straight-leg position
    Knee
        Extension
            Knee extensors (quadriceps)
                Rectus femoris
                Vastus lateralis
                Vastus medialis
                Vastus intermedius

Hip
    Extension
        Hip extensors
            Gluteus maximus
            Biceps femoris
            Semitendinosus
            Semimembranosus
Return to starting position
    Knee
        Flexion
            Knee extensors (eccentric contraction)
    Hip
        Flexion
            Hip extensors (eccentric contraction)

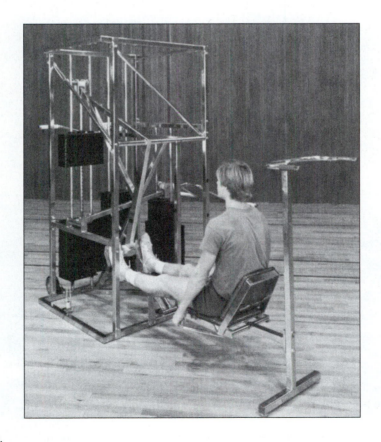

FIG. 11.8 ● Leg press.

Hip sled   FIG. 11.9

Description

The participant lies in a supine position on the floor with the knees and hips flexed in a position close to the chest. The feet are placed on the apparatus plate. The plate is moved upward until the knees and hips are completely extended. Then the participant returns to the starting position.

Analysis

This exercise is divided into two movements for analysis: (1) movement upward to high position and (2) return movement to the starting position.

Movement upward to high position
    Foot and ankle
        Plantar flexion
            Gastrocnemius
            Soleus
    Knee
        Extension
            Quadriceps
                Rectus femoris
                Vastus medialis
                Vastus intermedius
                Vastus lateralis

Hip
    Extension
        Biceps femoris
        Semimembranosus
        Semitendinosus
        Gluteus maximus
Return movement to starting position
    Foot and ankle
        Dorsal flexion
            Plantar flexors (eccentric contraction)
    Knee
        Flexion
            Knee extensors (eccentric contraction)
    Hip
        Flexion
            Hip extensors (eccentric contraction)

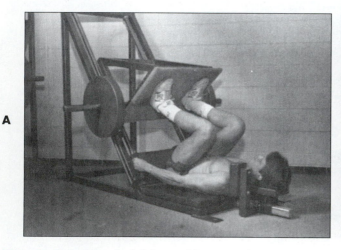
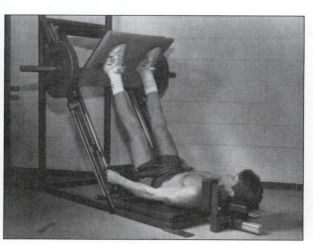

A   B

FIG. 11.9 ● Hip sled (hip and leg press). **A**, Starting position; **B**, high position.

Rowing exercise   FIG. 11.10

Description

The participant sits on a movable seat with the knees and hips flexed close to the chest. The arms are reaching forward to grasp a horizontal bar. The legs are extended forcibly as the arms are pulled toward the chest. Then the legs and arms are returned to the starting position.

Analysis

This exercise is divided into two movements for analysis: (1) movement to extend the legs forward and arms pulled toward the chest and (2) return movement to the starting position.

Movement to extend the legs forward and arms
  pulled toward the chest
    Foot and ankle
      Plantar flexion
        Gastrocnemius
        Soleus
    Knee
      Extension
        Rectus femoris
        Vastus intermedius
        Vastus lateralis
        Vastus medialis
    Hip
      Extension
        Gluteus maximus
        Biceps femoris
        Semimembranosus
        Semitendinosus
    Trunk
      Extension
        Erector spinae
    Shoulder girdle
      Adduction, downward rotation, and
        depression
        Shoulder girdle adductors, downward
          rotators, and depressors
            Trapezius (lower)
            Rhomboid
            Pectoralis minor
    Shoulder joint
      Extension
        Shoulder joint extensors
          Latissimus dorsi
          Teres major
          Posterior deltoid
          Teres minor
          Infraspinatus
    Elbow joint
      Flexion

      Elbow flexors
        Biceps brachii
        Brachialis
        Brachioradialis
    Wrist and hand
      Flexion
        Wrist and hand flexors (isometric con-
          traction)
          Flexor carpi radialis
          Flexor carpi ulnaris
          Palmaris longus
          Flexor digitorum profundus
          Flexor digitorum superficialis
          Flexor pollicis longus
Return movement to the starting position
    Foot and ankle
      Dorsal flexion
        Tibial anterior
        Extensor hallucis longus
        Extensor digitorum longus
        Peroneus tertius
    Knee
      Flexion
        Biceps femoris
        Semitendinosus
        Semimembranosus
    Hip
      Flexion
        Iliopsoas
        Rectus femoris
        Pectineus
    Trunk
      Flexion
        Rectus abdominis
        Internal oblique abdominal
        External oblique abdominal
    Shoulder girdle
      Abduction, upward rotation, and elevation
        Shoulder girdle adductors, downward
          rotators, and depressors (eccentric
          contraction)
    Shoulder joint
      Flexion
        Shoulder joint extensors (eccentric
          contraction)
    Elbow joint
      Extension
        Elbow joint flexors (eccentric
          contraction)
    Wrist and hand
      Flexion
        Wrist and hand flexors (isometric
          contraction)

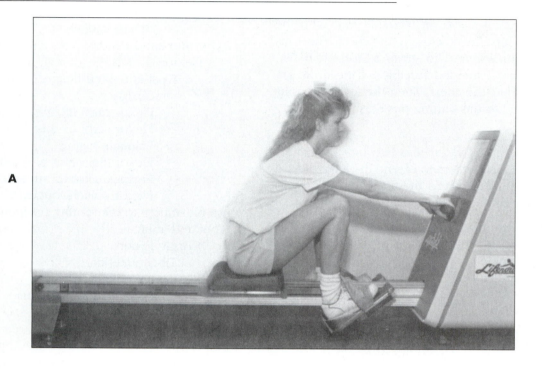

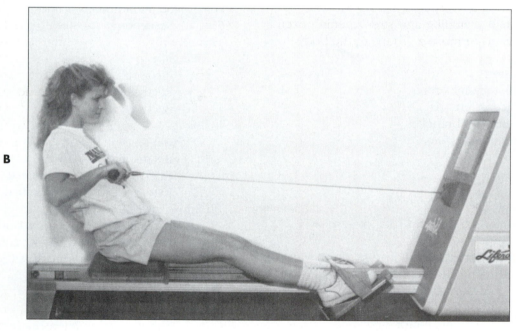

Web sites

Stretching and Flexibility: Everything you never wanted to know:

www.cs.huji.ac.il/papers/rma/stretching_toc.html

This paper by Brad Appleton gives detailed information on stretching and stretching techniques. It includes normal ranges of motion, flexibility, how to stretch, the physiology of stretching, and the types of stretching, including PNF.

Fitness World:

www.fitnessworld.com

The information at this site is about fitness in general and includes access to *Fitness Management* magazine.

Concept II:

www.concept2.com/index.html

Information on the technique of rowing and the muscles used.

Pump It Up!:

http://www.netspace.org/~gch/cs92/home.html

A site that has stretching and strengthening exercises for the major muscle groups of the body

Worksheet exercises

As an aid to learning, for in-class and out-of-class assignments, or for testing, a tear-out worksheet is found at the end of the text (pp. 267).

Skill analysis worksheets (no. 1 and no. 2)

Using the techniques taught in this chapter and Chapter 6, analyze the joint movements and muscles used in each phase of movement for selected exercises from the Total Gym examples on page 226 and/or from selected sports skills. For each joint, list the initial and subsequent position, the motion and degrees of movement, and the agonists and denote whether they are contracting concentrically or eccentrically.

Laboratory and review exercises

1. Obtain, describe, and completely analyze five conditioning exercises.
2. Collect, analyze, and evaluate exercises that are found in newspapers and magazines or are observed on television.
3. Prepare a set of exercises that will ensure development of all large muscle groups in the body.

4. Select exercises from exercise books for analysis.
5. Bring to class other typical exercises for members to analyze.
6. Analyze the conditioning exercises given by your physical education teachers, coaches, and athletic trainers.
7. Observe children using playground equipment. Analyze muscularly the activities they are performing.
8. Visit the room on your campus where the free weights and specific or multifunction exercise machines are located. Analyze exercises that can be done with each machine. Compare and contrast similar exercises using different exercise machines and free weights. NOTE: Manufacturers of all types of exercise apparatus have a complete list of exercises that can be performed with their machines. Secure a copy of recommended exercises and muscularly analyze each exercise.
9. Consider a sport (basketball or any other sport) and develop exercises applying the overload principle that would develop all the large muscle groups used in the sport.
10. Lie supine on a table with the knees flexed and hips flexed 90 degrees and the ankles in the neutral 90 degree. Push your foot straight away from the hip and knee flexed position until your knee is fully extended, your hip is flexed only 10 degrees, and your ankle is plantar flexed 10 degrees by performing each of the following movements before proceeding to the next:
    - Hip extension to within 10 degrees of flexion
    - Full knee extension
    - Plantar flexion to 10 degrees

    Analyze the movements and the muscles responsible for each movement at the hip, knee, and ankle.
11. Now stand with your back and buttocks against a smooth wall and place your feet (shoulder-width apart) with approximately 12 inches between your heels and the wall. Maintain your feet in position, with the hips and knees each flexed approximately 90 degrees so that the thighs are parallel to the floor. Keeping your feet in place, slowly slide your back and buttocks up the wall until your buttocks are as far away from the floor as possible without removing your feet from their position on the floor. Analyze the movements and the muscles responsible for each movement at the hip, knee, and ankle.

12. What is the difference between the two exercises in questions #10 and #11? Can you perform #11 one step at a time, as you did in #10?
13. Analyze each exercise in the exercise analysis chart. Use one row for each joint involved that actively moves during the exercise. Do not include joints where there is no active movement or where the joint is maintained in one position isometrically.

Exercise analysis chart

Exercise	Phase	Joint, movement occurring	Force causing movement (muscle or gravity)	Force resisting movement (muscle or gravity)	Functional muscle group, type of contraction
Sit-up—bent knee	Curling upward				
	Return movement				
Prone arch	Upward movement				
	Return movement				
Squat	Downward movement				
	Return to standing				
Dead lift	Upward movement				
	Return movement				
Leg press	Move to straight leg				
	Return movement				

Exercise analysis chart (*continued*)

Exercise	Phase	Joint, movement occurring	Force causing movement (muscle or gravity)	Force resisting movement (muscle or gravity)	Functional muscle group, type of contraction
Hip sled	Upward movement				
	Return movement				
Rowing exercise	Downward movement behind shoulders				
	Return movement				

References

Adrian M: Isokinetic exercise, *Training and Conditioning* 1:1, June 1991.

Altug Z, Hoffman JL, Martin JL: *Manual of clinical exercise testing, prescription and rehabilitation*, Norwalk, CT, 1993, Appleton & Lange.

Andrews JR, Harrelson GL: *Physical rehabilitation of the injured athlete*, Philadelphia, 1991, Saunders.

Baitch SP: Aerobic dance injuries, a biomechanical approach, *Journal of Physical Education, Recreation and Dance* 58:57, May–June 1987.

Bouche J: Three essential lifts for high school players, *Scholastic Coach* 56:42, April 1987.

Brzycki M: Rx for a safe productive strength program, *Scholastic Coach* 57:70, September 1987.

Epley B: Getting elementary muscles, *Coach and Athlete* 44:60, November-December 1981.

Fahey TD: *Athletic training: principles and practices*, Mountain View, CA, 1986, Mayfield.

Logan GA, McKinney WC: *Anatomic kinesiology*, ed 3, 1982, McGraw-Hill Companies, Inc., New York.

Matheson O, et al: Stress fractures in athletes, *American Journal of Sports Medicine* 15:46, January-February 1987.

Minton S: Dance dynamics avoiding dance injuries (symposium), *Journal of Physical Education, Recreation and Dance* 58:29, May–June 1987.

Northrip JW, Logan GA, McKinney WC: *Analysis of sport motion: anatomic and biomechanic perspectives*, ed 3, 1983, McGraw-Hill Companies, Inc., New York.

Prentice WE: *Rehabilitation techniques in sports medicine*, ed 3, 1999, McGraw-Hill Companies, Inc., New York.

Schlitz J: The athlete's daily dozen stretches, *Athletic Journal* 66:20, November 1985.

Steindler A: *Kinesiology of the human body*, Springfield, IL, 1970, Charles C Thomas.

Todd J: Strength training for female athletes, *Journal of Physical Education, Recreation and Dance*, 56:38, August 1985.

Torg JS, Vegso JJ, Torg E: *Rehabilitation of athletic injuries: an atlas of therapeutic exercise*, Chicago, 1987, Year Book.

Wirhed R: *Athletic ability and the anatomy of motion*, London, 1984, Wolfe.

Basic biomechanical factors and concepts

12 . . . .

Objectives

- To know and understand how knowledge of levers can help improve physical performance

- To know and understand how knowledge of torque, lever arm lengths, and angles of pull can help improve physical performance

- To know and understand how knowledge of Newton's laws of motion can help improve physical performance

- To know and understand how knowledge of balance, equilibrium, and stability can help improve physical performance

- To know and understand how knowledge of force and momentum can help improve physical performance

Many students in kinesiology classes have some knowledge, from a college or high school physics or physical science course, of the physical laws that affect motion. They need to review these facts and principles as they learn to apply them to motion in the human body. This area, the study of the mechanics as it relates to the functional and anatomical analysis of biological systems and especially humans, is known as *biomechanics*. Movements of the human body can be understood more completely by studying its mechanical characteristics and principles.

A brief discussion of some of these biomechanical characteristics and principles is included in this chapter.

Mechanics, the study of physical actions of forces, can be subdivided into *statics* and *dynamics*. Statics involves the study of systems that are in a constant state of motion, whether at rest with no motion or moving at a constant velocity without acceleration. Statics involves all forces acting on the body being in balance, resulting in the body being in equilibrium. Dynamics involves the study of systems in motion with acceleration. A system in acceleration is unbalanced due to unequal forces acting on the body. Additional components of biomechanical study include *kinematics* and *kinetics*. Kinematics is concerned with the description of motion and includes consideration of time, displacement, velocity, acceleration, and space factors of a system's motion. Kinetics is the study of forces associated with the motion of a body.

Levers

It is difficult for a person to visualize his or her body as a system of levers. The topic may seem academic to some, but this is far from true. A person moves through the use of his or her system of levers. The anatomical levers of the body cannot be changed, but, when properly understood, the system can be used more efficiently to maximize the muscular efforts of the body.

A lever is defined as a rigid bar that turns about an *axis* of rotation, or fulcrum. The axis is the point of rotation about which the lever moves.

The lever rotates about the axis as a result of *force* (sometimes referred to as effort, *E*) being applied to it to cause its movement against a *resistance* or weight. In the body, the bones represent the bars, the joints are the axes, and the muscles contract to apply the force. The amount of resistance can vary from maximal to minimal. In fact, the bones themselves or weight of the body segment may be the only resistance applied. All lever systems have each of these three components in one of three possible arrangements.

The arrangement or location of these three points in relation to one another determines the type of lever and for which kind of motion they are best suited. These points are the axis, the point of force application (usually the muscle insertion), and the point of resistance application (sometimes the center of gravity of the lever and sometimes the location of an external resistance). When the axis (*A*) is placed between the force (*F*) and the resistance (*R*), a first-class lever is produced (Figs. 12.1 and 12.2). In second-class levers, the resistance is between the axis and the force (Figs. 12.1 and 12.3). If the force is placed between the axis and the resistance, a third-class lever is created (Figs. 12.1 and 12.4).

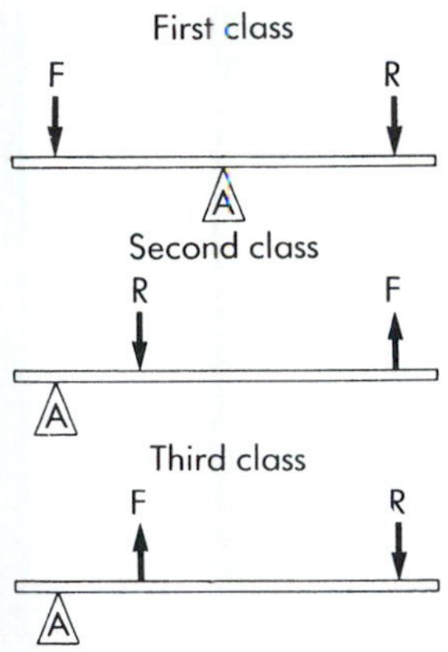

First-class levers

Typical examples of a first-class lever are the crowbar, seesaw, and elbow extension. An example of this type of lever in the body is seen with the triceps applying the force to the olecranon (*F*) in extending the nonsupported forearm (*R*) at the elbow (*A*). Other examples of this type of lever may be seen in the body when the agonist and antagonist muscle groups on either side of a joint axis are contracting simultaneously, with the agonist producing force and the antagonist supplying the resistance. A first-class lever (Fig. 12.2) is designed basically to produce balanced movements when the axis is midway between the force and the resistance (e.g., a seesaw). When the axis is close to the force, the lever produces speed and range of motion (e.g., the triceps in elbow extension). When the axis is close to the resistance, the lever produces force motion (e.g., a crowbar).

In applying the principle of levers to the body, it is important to remember that the force is applied where the muscle inserts in the bone, not in the belly of the muscle. For example, in elbow extension with the shoulder fully flexed and the arm beside the ear, the triceps applies the force to the olecranon of the ulna behind the axis of the elbow joint. As the applied force exceeds the amount of forearm resistance, the elbow extends.

The type of lever may be changed for a given joint and muscle, depending on whether the body segment is in contact with a surface such as a floor or wall. For example, we have demonstrated the triceps in elbow extension being a first-class lever with the hand free in space where the arm is pushed away from the body. By placing the hand in contact with the floor, as in performing a push-up to push the body away from the floor, the same muscle action at this joint now changes the lever to second class because the axis is at the hand and the resistance is the body weight at the elbow joint.

FIG. 12.1 ● Classification of levers.

Modified from Hall SJ: *Basic biomechanics*, St. Louis, 1991, Mosby.

FIG. 12.2 • **A** and **B**, First-class levers.

A modified from Booher JM, Thibodeau GA: *Athletic injury assessment*, ed 2, St., Louis, 1989, Mosby; **B** modified from Hall SJ, *Basic biomechanics*, St. Louis, 1991, Mosby.

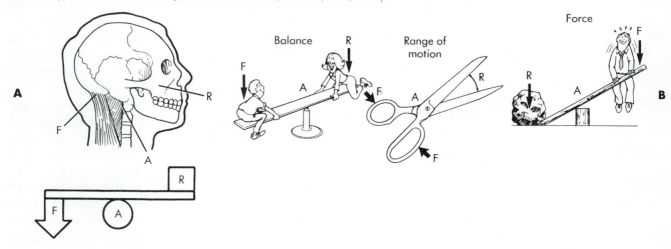

Second-class levers

A second-class lever (Fig. 12.3) is designed to produce force movements, since a large resistance can be moved by a relatively small force. An example of a second-class lever is a wheelbarrow. Besides the example given previously of the triceps extending the elbow in a push-up, a similar example of a second-class lever in the body is plantar flexion of the foot to raise the body up on the toes. The ball (*A*) of the foot serves as the axis of rotation as the ankle plantar flexors apply force to the calcaneus (*F*) to lift the resistance of the body at the tibial articulation (*R*) with the foot. There are relatively few occurrences of second-class levers in the body.

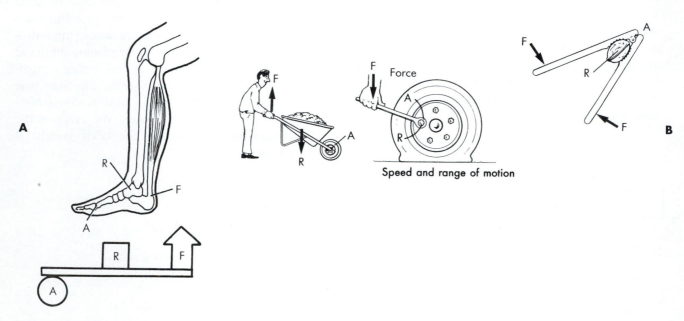

Speed and range of motion

FIG. 12.3 • **A** and **B**, Second-class levers.

A modified from Booher JM, Thibodeau GA: *Athletic injury assessment*, ed 2, St. Louis, 1989, Mosby; **B** modified from Hall SJ: *Basic biomechanics*, St. Louis, 1991, Mosby.

Third-class levers

Third-class levers (Fig. 12.4), with the force being applied between the axis and the resistance, are designed to produce speed and range-of-motion movements. Most of the levers in the human body are of this type, which require a great deal of force to move even a small resistance. Examples include a screen door operated by a short spring and application of lifting force to a shovel handle with the lower hand while the upper hand on the shovel handle serves as the axis of rotation. The biceps brachii is a typical example in the body. Using the elbow joint (*A*) as the axis, the biceps brachii applies force at its insertion on the radial tuberosity (*F*) to rotate the forearm up, with its center of gravity (*R*) serving as the point of resistance application.

The brachialis is an example of true third-class leverage. It pulls on the ulna just below the elbow, and, since the ulna cannot rotate, the pull is direct and true. The biceps brachii, on the other hand, supinates the forearm as it flexes, so that the third-class leverage applies to flexion only.

Other examples include the hamstrings contracting to flex the leg at the knee while in a standing position and using the iliopsoas to flex the thigh at the hip.

FIG. 12.4 ● **A** and **B**, Third-class levers.

A modified from Booher JM, Thibodeau GA: *Athletic injury assessment,* ed 2, St. Louis, 1989, Mosby; **B** modified from Hall SJ: *Basic biomechanics,* St. Louis, 1991, Mosby.

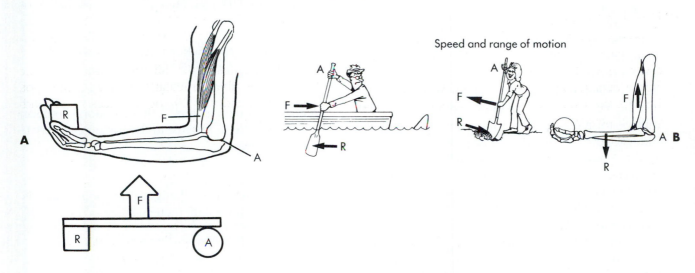

Factors in use of anatomical levers

Our anatomical leverage system can be used to gain a mechanical advantage that will improve simple or complex physical movements. Some individuals unconsciously develop habits of using human levers properly, but frequently this is not true.

Torque and length of lever arms

To understand the leverage system the concept of torque must be understood. *Torque*, or movement of force, is the turning effect of an eccentric force. *Eccentric force* is a force that is applied in a direction not in line with the center of rotation of an object with a fixed axis. In objects without a fixed axis, it is an applied force that is not in line with the object's center of gravity. For rotation to occur, an eccentric force must be applied. In the human body, the contracting muscle applies an eccentric force (not to be confused with eccentric contraction) to the bone on which it attaches and causes the bone to rotate about an axis at the joint. The amount of torque can be determined by multiplying the amount of force (*force magnitude*) by the *force arm*. The perpendicular distance between the location of force application and the axis is known as the force arm, moment arm, or torque arm. The force arm may be best understood by realizing it is the shortest distance from the axis of rotation to the line of action of the force. The greater the distance of the force arm, the more torque produced by the force. A practical application of torque and levers often used is when we purposely increase the force arm length in order to increase the torque so that we can more easily move a relatively large resistance. This is commonly referred to as increasing our leverage.

It is also important to note the *resistance arm*, which may be defined as the distance between the axis and the point of resistance application. In discussing the application of levers, it is necessary to understand the length relationship between the two lever arms. There is an inverse relationship between force and the force arm, just as there is between resistance and the resistance arm. The longer the force arm, the less force required to move the lever if the resistance and resistance arm remain constant. In addition, if the force and force arm remain constant, a greater resistance may be moved by shortening the resistance arm.

Also, there is a proportional relationship between the force components and the resistance components. That is, for movement to occur when either of the resistance components increases, there must be an increase in one or both of the force components. Even slight variations in the location of the force and resistance are important in determining the effective force of the muscle. This point can be illustrated in the following simple formula, using the biceps brachii muscle in each example:

$$F \times FA = R \times RA$$

(Force) × (Force = (Resistance) × (Resistance arm) arm)

Initial example

$$F \times 0.1 \text{ meter} = 45 \text{ Newtons} \times 0.25 \text{ meter}$$
$$0.1 \times F = 11.25 \text{ Newton-meters}$$
$$F = 112.5 \text{ Newton-meters}$$

Example A

Increase the (FA) by changing the insertion 0.05 meters:

$$F \times 0.15 \text{ meter} = 45 \text{ Newtons} \times 0.25 \text{ meter}$$
$$0.15 \times F = 11.25 \text{ Newton-meters}$$
$$F = 75 \text{ Newton-meters}$$

A change of 0.05 meters in the insertion can make a considerable difference in the force necessary to move the lever.

Example B

Reduce the (RA) by changing the point of resistance application 0.05 meters:

$$F \times 0.1 \text{ meter} = 45 \text{ Newtons} \times 0.2 \text{ meter}$$
$$0.1 \times F = 9 \text{ Newton-meters}$$
$$F = 90 \text{ Newton-meters}$$

Shortening the resistance arm can decrease the amount of force necessary to move the lever.

Example C

Change the (R) amount by reducing the resistance 1 Newton:

$$F \times 0.1 \text{ meter} = 44 \text{ Newtons} \times 0.25 \text{ meter}$$
$$0.1 \times F = 11 \text{ Newton-meters}$$
$$F = 110 \text{ Newton-meters}$$

Obviously, decreasing the amount of resistance can decrease the amount of force needed to move the lever.

The system of leverage in the human body is built for speed and range of movement at the expense of force. Short force arms and long resistance arms require great muscular strength to produce movement. In the forearm, the attachments of the biceps and triceps muscles clearly illustrate this point, since the force arm of the biceps is 1 to 2 inches and that of the triceps less than 1 inch. Many other similar examples are found all over the body. From a practical point of view, this means that the muscular system should be strong to supply the necessary force for body movements, especially in strenuous sports activities.

When we speak of human leverage in relation to sport skills, we are generally referring to several levers. For example, in throwing a ball, there are levers at the shoulder, elbow, and wrist joints.

In fact, it can be said that there is one long lever from the feet to the hand.

The longer the lever, the more effective it is in imparting velocity. A tennis player can hit a tennis ball harder with a straight-arm drive than with a bent elbow because the lever (including the racket) is longer and moves at a faster speed.

Fig. 12.5 indicates that a longer lever (Z') travels faster than a shorter lever (S') in traveling the same number of degrees. In sports activities in which it is possible to increase the length of a lever with a racket or bat, the same principle applies.

In baseball, hockey, golf, field hockey, and other sports, long levers similarly produce more linear force and thus better performance. For quickness of movement, it is sometimes desirable to have a short lever arm, such as when a baseball catcher brings the hand back to the ear to secure a quick throw or when a sprinter shortens the knee lever through flexion so much that the sprinter almost catches his or her spikes in his or her gluteal muscles.

FIG. 12.5 ● Length of levers.

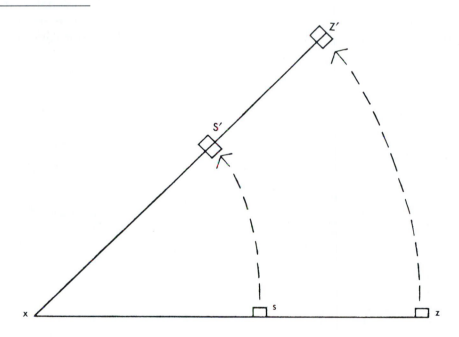

Angle of pull

Another factor of considerable importance in using the leverage system is the angle of pull of the muscles on the bone. The angle of pull may be defined as the angle between the line of pull of the muscle and the bone on which it inserts. For the sake of clarity and consistency, we need to specify that the actual angle referred to is the angle of attachment facing away from the joint as opposed to the angle on the side of the joint. With every degree of joint motion, the angle of pull changes. Joint movements and insertion angles involve mostly small angles of pull. The angle of pull decreases as the bone moves away from its anatomical position through the contraction of the local muscle group. This range of movement depends on the type of joint and bony structure.

Most muscles work at a small angle of pull, generally less than 50 degrees. The amount of muscular force needed to cause joint movement is affected by the angle of pull. Three components of muscular force are involved. The *rotary component*, also referred to as the vertical component, is the component of muscular force that acts perpendicular to the long axis of the bone (lever). When the line of muscular force is at 90 degrees to the bone on which it attaches, all of the muscular force is rotary force; therefore, 100% of the force is contributing to the movement. That is, all of the force is being used to rotate the lever about its axis. The closer the angle of pull is to 90 degrees, the greater the rotary component. At all other degrees of the angle of pull, one of the other two components of force are operating in addition to the rotary component. The same rotary component is continuing, although with less force, to rotate the lever about its axis. The second force component is the horizontal, or *nonrotary component* and is either a *stabilizing component* or a *dislocating component*, depending on whether the angle of pull is less than or greater than 90 degrees. If the angle is less than 90 degrees, the force is a stabilizing force because its pull directs the bone toward the joint axis. If the angle is greater than 90 degrees, the force is dislocating due to its pull directing the bone away from the joint axis (Fig. 12.6).

In some activities, it is desirable to have a person begin a movement when the angle of pull is at 90 degrees. Many boys and girls are unable to do a chin-up (pull-up) unless they start with the elbow in a position to allow the elbow flexor muscle group to approximate a 90-degree angle with the forearm.

FIG. 12.6 ● **A** to **C**, Components of force due to the angle of pull.

Modified from Hall SJ: *Basic biomechanics*, St. Louis, 1991, Mosby.

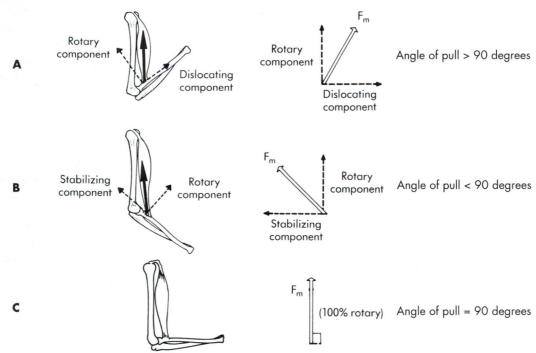

This angle makes the chin-up easier because of the more advantageous angle of pull. The application of this fact can compensate for lack of sufficient strength. In its range of motion, a muscle pulls a lever through a range characteristic of itself, but, in approaching and going beyond 90 degrees, it is most effective. An increase in strength is the only solution for muscles that operate at disadvantageous angles of pull and require a greater force to operate efficiently.

Laws of motion and physical activities

Motion is fundamental in physical education and sports activity. Body motion is generally produced or at least started by some action of the muscular system. Motion cannot occur without a force, and the muscular system is the source of force in the human body. Thus, development of the muscular system is indispensable to movement.

Basically, there are two types of motion: *linear motion* and *angular motion*. Linear motion, also referred to as translatory motion, is motion along a line. If the motion is along a straight line, it is *rectilinear* motion, whereas motion along a curved line is known as *curvilinear* motion. Angular motion, also known as rotary motion, involves rotation around an axis. In the human body, the axis of rotation is provided by the various joints. In a sense, these two types of motion are related, since angular motion of the joints can produce the linear motion of walking.

For example, in many sports activities, the cumulative angular motion of the joints of the body imparts linear motion to a thrown object (ball, shot) or to an object struck with an instrument (bat, racket).

Displacement refers to a change in position or location of an object from its original point of reference, whereas *distance*, or the path of movement, refers to the actual sum length of measurement traveled. An object may have traveled a distance of 10 meters along a linear path in two or more directions but be displaced from its original reference point by only 6 meters. *Angular displacement* refers to the change in location of a rotating body. *Linear displacement* is the distance that a system moves in a straight line.

We are sometimes concerned about the time it takes for displacement to occur. *Speed* is how fast an object is moving, or the distance an object travels in a specific amount of time. *Velocity* includes the direction and describes the rate of displacement.

A brief review of Newton's laws of motion will indicate the many applications of these laws to physical education activities and sports. Newton's laws explain all the characteristics of motion, and they are fundamental to understanding human movement.

Law of inertia

A body in motion tends to remain in motion at the same speed in a straight line unless acted on by a force; a body at rest tends to remain at rest unless acted on by a force.

Inertia may be described as the resistance to action or change. In terms of human movement, *inertia* refers to resistance to acceleration or deceleration. Inertia is the tendency for the current state of motion to be maintained, regardless of whether the body segment is moving at a particular velocity or is motionless.

Muscles produce the force necessary to start motion, stop motion, accelerate motion, decelerate motion, or change the direction of motion. Put another way, inertia is the reluctance to change status; only force can do it. The greater the mass of an object, the greater its inertia. Therefore, the greater the mass, the more force needed to significantly change an object's inertia. Numerous examples of this law are found in physical education activities. A sprinter in the starting blocks must apply considerable force to overcome resting inertia. A runner on an indoor track must apply considerable force to overcome moving inertia and stop before hitting the wall. Balls and other objects that are thrown or struck require force to stop them. Starting, stopping, and changing direction—a part of many physical activities—provide many other examples of the law of inertia applied to body motion.

Since force is required to change inertia, it is obvious that any activity that is carried out at a steady pace in a consistent direction will conserve energy and that any irregularly paced or directed activity will be very costly to energy reserves. This explains in part why activities such as handball and basketball are so much more fatiguing than jogging or dancing.

Law of acceleration

A change in the acceleration of a body occurs in the same direction as the force that caused it. The change in acceleration is directly proportional to the force causing it and inversely proportional to the mass of the body.

Acceleration may be defined as the rate of change in velocity. To attain speed in moving the body, a strong muscular force is generally necessary. *Mass*, the amount of matter in a body, affects the speed and acceleration in physical movements. A much greater force is required from the muscles to accelerate an 80-kilogram man than to accelerate a 58-kilogram man to the same running speed. Also, it is possible to accelerate a baseball faster than a shot because of the difference in weight. The force required to run at half speed is less than the force required to run at top speed. To impart speed to a ball or an object, it is necessary to rapidly accelerate the part of the body holding the object. Football, basketball, track, and field hockey are a few sports that demand speed and acceleration.

Law of reaction

For every action there is an opposite and equal reaction.

As we place force on a supporting surface by walking over it, the surface provides an equal resistance back in the opposite direction to the soles of our feet. Our feet push down and back, while the surface pushes up and forward. The force of the surface reacting to the force we place on it is referred to as *ground reaction force*. We provide the action force, while the surface provides the reaction force. It is easier to run on a hard track than on a sandy beach because of the difference in the ground reaction forces of the two surfaces. The track resists the runner's propulsion force, and the reaction drives the runner ahead. The sand dissipates the runner's force, and the reaction force is correspondingly reduced with the apparent loss in forward force and speed. A sprinter applies a force in excess of 1335 Newtons on the starting blocks, which resist with an equal force. When a body is in flight, as it is in jumping, movement of one part of the body produces a reaction in another part. This occurs because there is no resistive surface to supply a reaction force.

Balance, equilibrium, and stability

Balance is the ability to control equilibrium, either static or dynamic. In relation to human movement, *equilibrium* refers to a state of zero acceleration where there is no change in the speed or direction of the body. Equilibrium may be either static or dynamic. If the body is at rest or completely motionless, it is in *static equilibrium. Dynamic equilibrium* occurs when all of the applied and inertial forces acting on the moving body are in balance, resulting in movement with unchanging speed or direction. For us to control equilibrium and, hence, achieve balance, we need to maximize *stability*. Stability is the resistance to a change in the body's acceleration or, more appropriately, the resistance to a disturbance of the body's equilibrium. Stability may be enhanced by determining the body's *center of gravity* and changing it appropriately. The center of gravity is the point at which all of the body's mass and weight is equally balanced or equally distributed in all directions.

Balance is important for the resting body, as well as for the moving body. Generally, balance is to be desired, but there are circumstances in which movement is improved when the body tends to be unbalanced. Following are certain general factors that apply toward enhancing equilibrium, maximizing stability, and ultimately achieving balance:

1. A person has balance when the center of gravity falls within the base of support.

2. A person has balance in direct proportion to the size of the base. The larger the base of support, the more balance.

3. A person has balance depending on the weight (mass). The greater the weight, the more balance.

4. A person has balance depending on the height of the center of gravity. The lower the center of gravity, the more balance.

5. A person has balance depending on where the center of gravity is in relation to the base of support. The balance is less if the center of gravity is near the edge of the base. However, when anticipating an oncoming force, stability may be improved by placing the center of gravity nearer the side of the base of support expected to receive the force.

6. In anticipation of an oncoming force, stability may be increased by enlarging the size of the base of support in the direction of the anticipated force.

7. Equilibrium may be enhanced by increasing the friction between the body and the surfaces it contacts.

8. Rotation about an axis aids balance. A moving bike is easier to balance than a stationary bike.

9. Kinesthetic physiological functions contribute to balance. The semicircular canals of the inner ear, vision, touch (pressure), and kinesthetic sense all provide balance information to the performer. Balance and its components of equilibrium and stability are essential in all movements. They are all affected by the constant force of gravity as well as by inertia. Walking has been described as an activity in which a person throws the body in and out of balance with each step. In rapid running movements in which moving inertia is high, the individual has to lower the center of gravity to maintain balance when stopping or changing direction. On the other hand, jumping activities attempt to raise the center of gravity as high as possible.

Force

Muscles are the main source of force that produces or changes movement of a body segment, the entire body, or an object thrown, struck, or stopped. Strong muscles are able to produce more force than weak muscles. This refers to both maximum and sustained exertion over a period of time.

Forces either push or pull on an object in an attempt to affect motion or shape. Without forces acting on an object, there is no motion. Force is the product of mass times acceleration. The mass of a body segment or the entire body times the speed of acceleration determines the force. Obviously, in football this is very important, yet it is just as important in other activities that use only a part of the human body. When one throws a ball, the force applied to the ball is equal to the mass of the arm times the arm's speed of acceleration. Also, as previously discussed, leverage is important.

$$\text{Force} = \text{mass} \times \text{acceleration}$$
$$F = m \times a$$

The quantity of motion or, more scientifically stated, the *momentum*, which is equal to mass times velocity, is important in skill activities. The greater the momentum, the greater the resistance to change in the inertia or state of motion.

It is not necessary to apply maximum force and thus increase the momentum of a ball or an object being struck in all situations. In skillful performance, regulation of the amount of force is necessary. Judgment as to the amount of force required to throw a softball a given distance, hit a golf ball 200 yards, or hit a tennis ball across the net and into the court is important.

In activities involving movement of various joints, as in throwing a ball or putting a shot, there should be a summation of forces from the beginning of movement in the lower segment of the body to the twisting of the trunk and movement at the shoulder, elbow, and wrist joints. The speed at which a golf club strikes the ball is the result of a summation of forces of the trunk, shoulders, arms, and wrists. Shot-putting and discus and javelin throwing are other good examples that show that summation of forces is essential.

Throwing

In the performance of various sport skills, many applications of the laws of leverage, motion, and balance may be found. A skill common to many activities is throwing. The object thrown may be some type of ball, but it is frequently an object of another size or shape, such as a rock, beanbag, Frisbee, discus, or javelin. A brief analysis of some of the basic mechanical principles involved in the skill of throwing will help indicate the importance of understanding the applications of these principles. Many activities involve these and sometimes other mechanical principles. Motion is basic to throwing when the angular motion (p. 240) of the levers (bones) of the body (trunk, shoulder, elbow, and wrist) is used to give linear motion to the ball when it is released.

Newton's laws of motion apply in throwing because the individual's inertia and the ball's inertia (p. 241) must be overcome by the application of force. The muscles of the body provide the force to move the body parts and the ball held in the hand. The *law of acceleration* (Newton's second law) comes into operation with the muscular force necessary to accelerate the arm, wrist, and hand. The greater the force (mass times acceleration) that a person can produce, the faster the arm will move and, thus, the greater the speed that will be imparted to the ball. The reaction of the feet against the surface on which the person stands indicates the application of the *law of reaction*.

The leverage factor is very important in throwing a ball or an object. The longer the lever, the greater the speed that can be imparted to the ball. For all practical purposes, the body from the feet to the fingers can be considered as one long lever. The longer the lever, either from natural body length or from the movements of the body to the extended backward position (as in throwing a softball, with extension of the shoulder and the elbow joints), the greater will be the arc through which it accelerates and thus the greater the speed imparted to the thrown object.

In certain circumstances, when the ball is to be thrown only a short distance, as in baseball when it is thrown by the catcher to the bases, the short lever is advantageous because it takes less total time to release the ball.

Balance, or equilibrium, is a factor in throwing when the body is rotated to the rear in the beginning of the throw. This motion moves the body nearly out of balance to the rear, and then balance changes again in the body with the forward movement. Balance is again established with the follow-through, when the feet are spread and the knees and trunk are flexed to lower the center of gravity.

Summary

The preceding discussion has been a brief overview of some of the factors affecting motion. Analysis of human motion in light of the laws of physics poses a problem: how comprehensive is the analysis to be? It can become very complex, particularly when body motion is combined with the manipulation of an object in the hand involved in throwing, kicking, striking, or catching.

These factors become involved when an analysis is attempted of the activities common to our physical education program—football, baseball, basketball, track and field, field hockey, and swimming, to mention a few. However, a physical educator who is to have a complete view of which factors control human movement must have a working knowledge of both the physiological and the biomechanical principles of kinesiology.

It is beyond the scope of this book to make a detailed analysis of other activities. Some sources that consider these problems in detail are listed in the references.

Web sites

Biomechanics World Wide:

www.per.ualberta.ca/biomechanics

This site enables the reader to search the biomechanics journals for recent information regarding mechanism of injury.

Biomechanics: The Magazine of Body Movement and Medicine:

www.biomech.com/

University of Arkansas Medical School Gross Anatomy for Medical Students:

anatomy.uams.edu/htmlpages/anatomyhtml/gross.html

Dissections, anatomy tables, atlas images, links, etc.

Worksheet exercises

1. Develop special projects and class reports by individual or small groups of students on the mechanical analysis of all the skills involved in the following:
   a. Basketball
   b. Baseball
   c. Dancing
   d. Diving
   e. Football
   f. Field hockey
   g. Golf
   h. Gymnastics
   i. Soccer
   j. Swimming
   k. Tennis
   l. Wrestling
2. Develop term projects and special class reports by individual or small groups of students about the following factors in motion:
   a. Balance
   b. Force
   c. Gravity
   d. Motion
   e. Torque
   f. Leverage
   g. Projectiles
   h. Friction
   i. Buoyancy
   j. Aerodynamics
   k. Hydrodynamics
   l. Restitution
   m. Spin
   n. Rebound angle
   o. Momentum
   p. Center of gravity
   q. Equilibrium
   r. Stability
   s. Base of support
   t. Inertia
   u. Linear displacement
   v. Angular displacement
   w. Speed
   x. Velocity
3. Develop demonstrations, term projects, or special reports by individual or small groups of students on the following activities:
   a. Lifting
   b. Throwing
   c. Standing
   d. Walking
   e. Running
   f. Jumping
   g. Falling
   h. Sitting
   i. Pushing and pulling
   j. Striking

References

Adrian MJ, Cooper JM: *The biomechanics of human movement,* Indianapolis, IN, 1989, Benchmark Press.

American Academy of Orthopaedic Surgeons: *Athletic training and sports medicine,* ed 2, Park Ridge, IL, 1991, American Academy of Orthopaedic Surgeons.

Barham JN: *Mechanical kinesiology,* St. Louis, 1978, Mosby.

Broer MR: *An introduction to kinesiology,* Englewood Cliffs, NJ, 1968, Prentice-Hall.

Broer MR, Zernicke RF: *Efficiency of human movement,* ed 3, Philadelphia, 1979, Saunders.

Bunn JW: *Scientific principles of coaching,* ed 2, Englewood Cliffs, NJ, 1972, Prentice-Hall.

Cooper JM, Adrian M, Glassow RB: *Kinesiology,* ed 5, St. Louis, 1982, Mosby.

Donatelli R, Wolf SL: *The biomechanics of the foot and ankle,* Philadelphia, 1990, Davis.

Hall SJ: *Basic biomechanics,* ed 4, 2000, McGraw-Hill Companies, Inc., New York.

Hinson M: *Kinesiology,* ed 2, 1981, McGraw-Hill Companies, Inc., New York.

Kegerreis S, Jenkins WL, Malone TR: Throwing injuries, *Sports Injury Management* 2:4, 1989.

Kelley DL: *Kinesiology: fundamentals of motion description,* Englewood Cliffs, NJ, 1971, Prentice-Hall.

Kreighbaum E, Barthels KM: *Biomechanics: a qualitative approach for studying human movement,* ed 3, New York, 1990, Macmillan.

Logan GA, McKinney WC: *Anatomic kinesiology,* ed 3, 1982, McGraw-Hill Companies, Inc., New York.

Luttgens K, Hamilton N: *Kinesiology: scientific basis of human motion,* ed 9, Madison, WI, 1997, Brown & Benchmark.

Nordin M, Frankel VH: *Basic biomechanics of the musculoskeletal system,* ed 2, Philadelphia, 1989, Lea & Febiger.

Norkin CC, Levangie PK: *Joint structure and function: a comprehensive analysis,* Philadelphia, 1983, Davis.

Northrip JW, Logan GA, McKinney WC: *Analysis of sport motion: anatomic and biomechanic perspectives,* ed 3, 1983, McGraw-Hill Companies, Inc., New York.

Piscopo J, Baley J: *Kinesiology: the science of movement,* New York, 1981, John Wiley & Sons.

Rasch PJ: *Kinesiology and applied anatomy,* ed 7, Philadelphia, 1989, Lea & Febiger.

Scott MG: *Analysis of human motion,* ed 2, New York, 1963, Appleton-Century-Crofts.

Weineck J: *Functional anatomy in sports,* ed 2, St. Louis, 1990, Mosby.

Wirhed R: *Athletic ability and the anatomy of motion,* London, 1984, Wolfe.

Appendix

Appendix 1

Range of motion for diarthrodial joints of the upper extremity

Joint	Type	Motion	Range
Sternoclavicular	Arthrodial	Protraction	Moves anteriorly 15°
		Retraction	Moves posteriorly 15°
		Elevation	Moves superiorly 45°
		Depression	Moves inferiorly 5°
Acromioclavicular	Arthrodial	Protraction-retraction	20–30° rotational and gliding motion
		Elevation-depression	20–30° rotational and gliding motion
		Upward rotation–downward rotation	20–30° rotational and gliding motion
Scapulothoracic	Not a true synovial joint, all movement totally dependent on AC and SC joints	Abduction-adduction	25° total range
		Upward rotation–downward rotation	60° total range
		Elevation-depression	55° total range
Glenohumeral	Enarthrodial	Flexion	90°–100°
		Extension	40°–60°
		Abduction	90°–95°
		Adduction	0° prevented by trunk, 75° anterior to trunk
		Internal rotation	70°–90°
		External rotation	70°–90°
		Horizontal abduction	45°
		Horizontal adduction	135°
Elbow	Ginglymus	Extension	0°
		Flexion	145°–150°
Radioulnar	Trochoid	Supination	80°–90°
		Pronation	70°–90°
Wrist	Condyloid	Flexion	70°–90°
		Extension	65°–85°
		Abduction	15°–25°
		Adduction	25°–40°

Range of motion for diarthrodial joints of the upper extremity (*continued*)

Joint	Type	Motion	Range
Thumb carpometacarpal	Sellar	Flexion	15°–45°
		Extension	0°–20°
		Adduction	0°
		Abduction	50°–70°
Thumb metacarpophalangeal	Ginglymus	Extension	0°
		Flexion	40°–90°
Thumb interphalangeal	Ginglymus	Flexion	80°–90°
		Extension	0°
2nd, 3rd, 4th, and 5th metacarpophalangeal joints	Condyloid	Extension	0°–40°
		Flexion	85°–100°
		Abduction	Variable 10°–40°
		Adduction	Variable 10°–40°
2nd, 3rd, 4th, and 5th proximal interphalangeal joints	Ginglymus	Flexion	90°–120°
		Extension	0°
2nd, 3rd, 4th, and 5th distal interphalangeal joints	Ginglymus	Flexion	80°–90°
		Extension	0°

Appendix 2

Range of motion for diarthrodial joints of the spine and lower extremity

Joint	Type	Motion	Range
Cervical	Arthrodial except atlantoaxial joint, which is trochoid	Flexion	80°
		Extension	20°–30°
		Lateral flexion	35°
		Rotation unilaterally	45°
Lumbar	Arthrodial	Flexion	45°
		Extension	45°
		Lateral flexion	45°
		Rotation unilaterally	60°
Hip	Enarthrodial	Flexion	130°
		Extension	30°
		Abduction	35°
		Adduction	0°–30°
		External rotation	50°
		Internal rotation	45°
Knee *For internal and external rotation to occur, the knee must be flexed 30° or more.*	Ginglymus	Extension	0°
		Flexion	140°
		Internal rotation	30°
		External rotation	45°
Ankle (talocrural)	Ginglymus	Plantar flexion	50°
		Dorsal flexion	15°–20°
Transverse tarsal and subtalar	Arthrodial	Inversion	20°–30°
		Eversion	5°–15°
Great toe metatarsophalangeal	Condyloid	Flexion	45°
		Extension	70°
		Abduction	Variable 5°–25°
		Adduction	Variable 5°–25°
Great toe interphalangeal	Ginglymus	Flexion	90°
		Extension	0°
2nd, 3rd, 4th, and 5th metatarsophalangeal joints	Condyloid	Flexion	40°
		Extension	40°
		Abduction	Variable 5°–25°
		Adduction	Variable 5°–25°
2nd, 3rd, 4th, and 5th proximal interphalangeal joints	Ginglymus	Flexion	35°
		Extension	0°
2nd, 3rd, 4th, and 5th distal interphalangeal joints	Ginglymus	Flexion	60°
		Extension	30°

Worksheets

Chapter one

Worksheet no. 1

On the posterior skeletal worksheet, list the names of the bones and all of the prominent features of each bone.

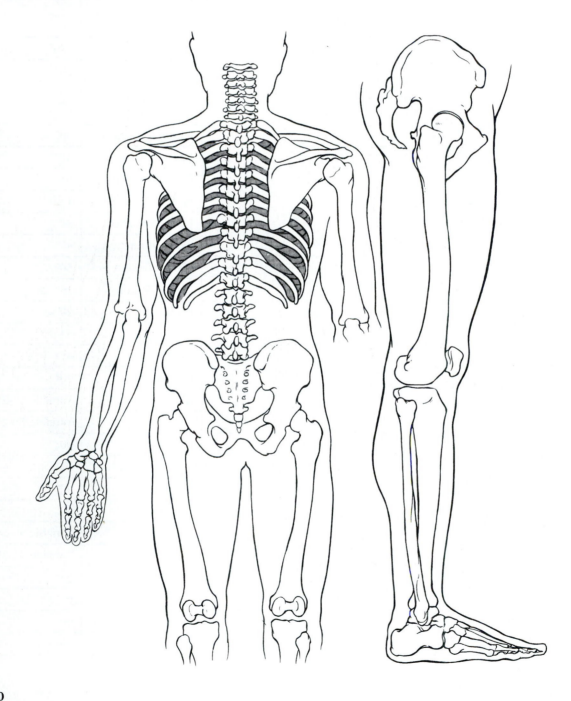

Chapter one

Worksheet no. 2

On the anterior skeletal worksheet, list the names of the bones and all of the prominent features of each bone.

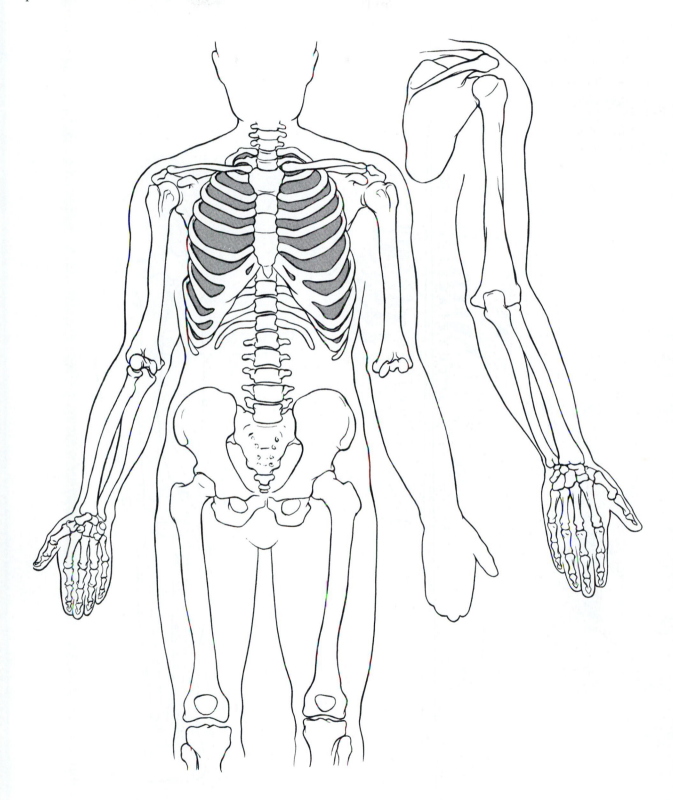

Chapter two

Worksheet no. 1

Draw and label on the worksheet the following listed muscles. Indicate the origin and insertion of each muscle with an "O" and "I," respectively.

a. Trapezius

b. Rhomboid

c. Serratus anterior

d. Levator scapulae

e. Pectoralis minor

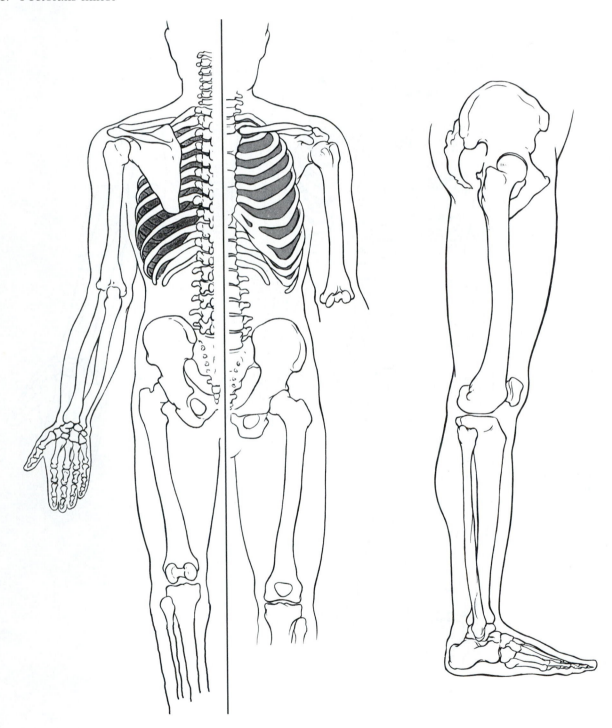

Chapter two

Worksheet no. 2

Label and indicate by arrows the following movements of the shoulder girdle. For each motion, list the plane in which it occurs and list the axis of rotation.

a. Adduction

b. Abduction

c. Upward rotation

d. Downward rotation

e. Elevation

f. Depression

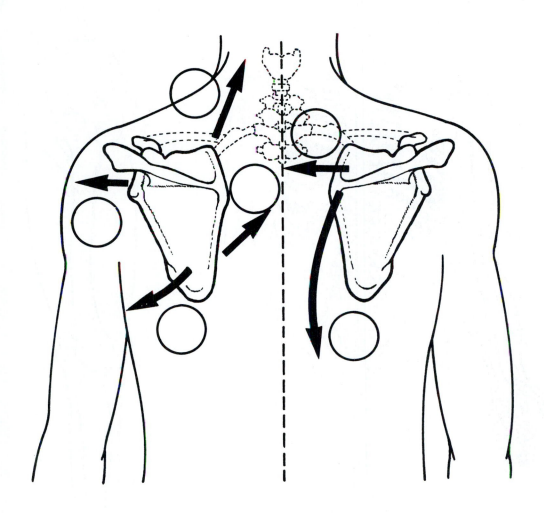

Chapter three

Worksheet no. 1

Draw and label on the worksheet the following muscles. Indicate the origin and insertion of each muscle with an "O" and "I," respectively.

a. Deltoid
b. Supraspinatus
c. Subscapularis
d. Teres major
e. Infraspinatus
f. Teres minor
g. Latissimus dorsi
h. Pectoralis major
i. Coracobrachialis

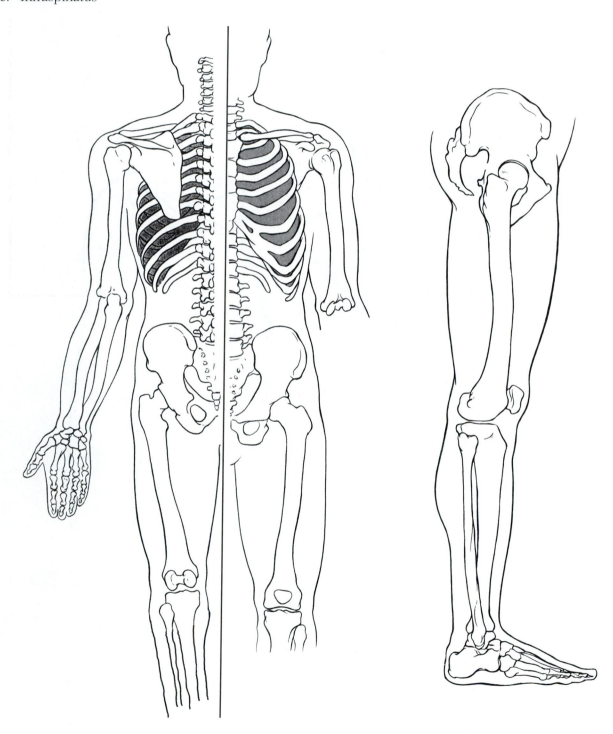

Chapter three

Worksheet no. 2

Label and indicate by arrows the following listed movements of the shoulder joint. For each motion, list the plane in which it occurs and list the axis of rotation.

a. Abduction   c. Flexion   e. Horizontal adduction

b. Adduction   d. Extension   f. Horizontal abduction

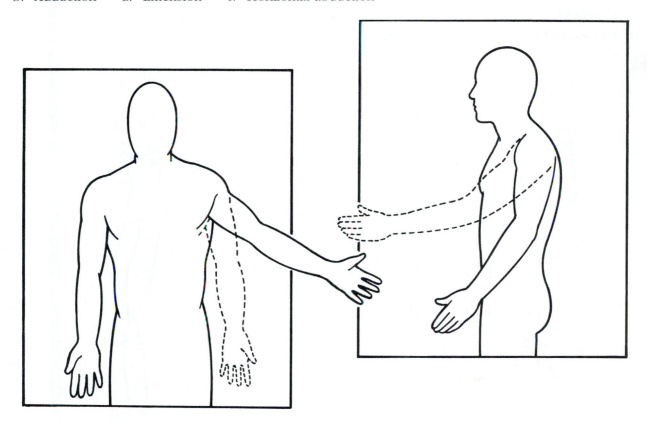

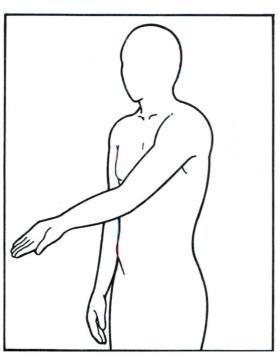

Chapter four

Worksheet no. 1

Draw and label on the worksheet the following muscles. Indicate the origin and insertion of each muscle with an "O" and "I," respectively.

a. Biceps brachii   e. Supinator

b. Brachioradialis   f. Triceps brachii

c. Brachialis   g. Anconeus

d. Pronator teres   h. Pronator quadratus

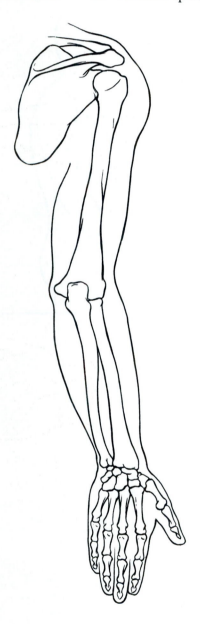
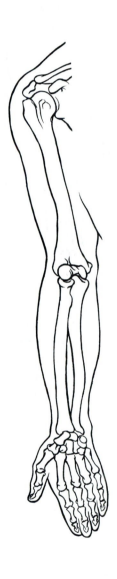

Chapter four

Worksheet no. 2

Label and indicate by arrows the following movements of the elbow and radioulnar joint. For each motion, list the plane in which it occurs and list the axis of rotation.

Elbow
  Flexion
  Extension

Radioulnar joint
  Pronation
  Supination

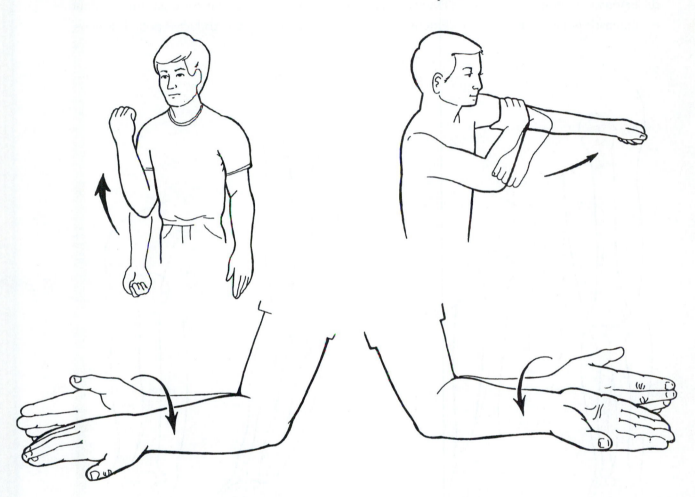

Chapter five

Worksheet no. 1

Draw and label on the worksheet the following muscles. Indicate the origin and insertion of each muscle with an "O" and "I," respectively.

a. Flexor pollicis longus
b. Flexor carpi radialis
c. Flexor carpi ulnaris
d. Extensor digitorum
e. Extensor pollicis longus

f. Extensor pollicis brevis
g. Extensor carpi ulnaris
h. Palmaris longus
i. Extensor carpi radialis longus
j. Extensor carpi radialis brevis

k. Extensor digiti minimi
l. Extensor digitorum indicis
m. Flexor digitorum superficialis
n. Flexor digitorum profundus
o. Abductor pollicis longus

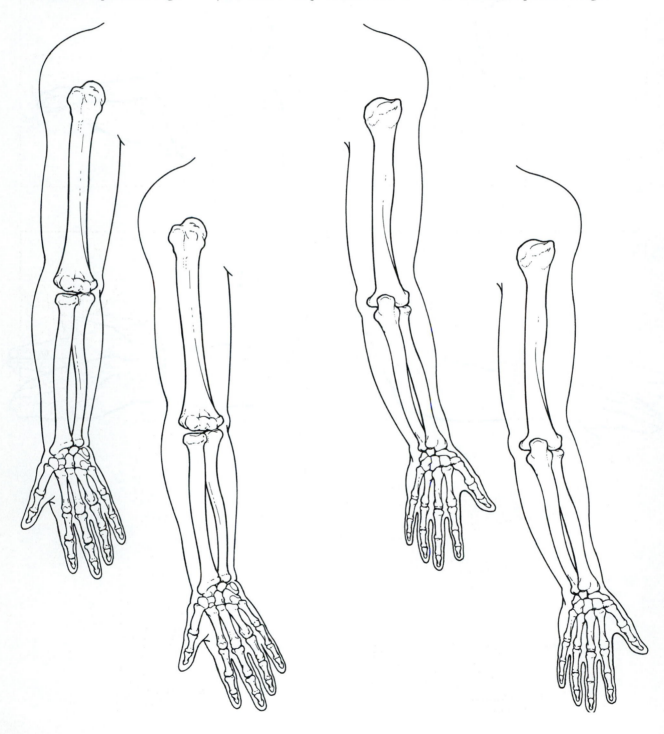

Chapter five

Worksheet no. 2

Label and indicate by arrows the following movements of the wrist and hands.
For each motion, list the plane in which it occurs and list the axis of rotation.

Wrist and hands
   Extension
   Flexion
   Abduction
   Adduction

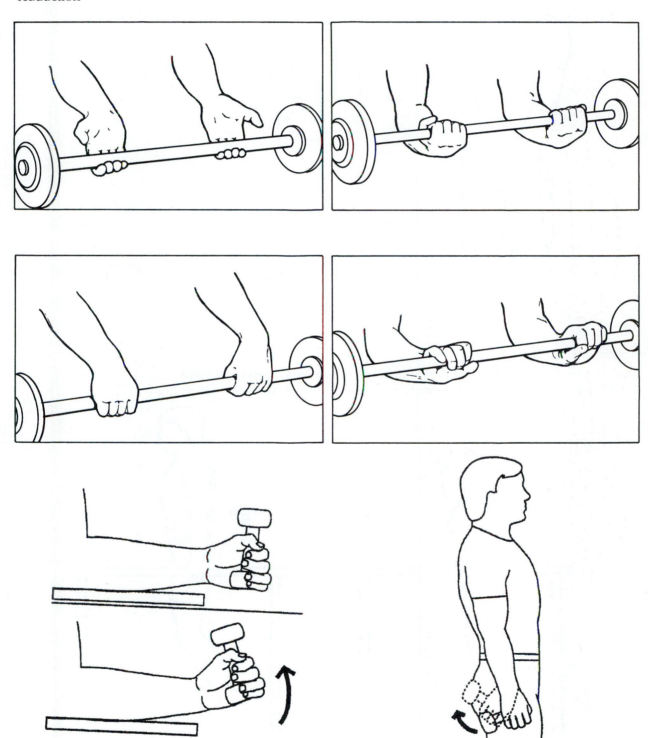

Chapter six

Worksheet no. 1

Analyze this exercise following the procedures explained in this chapter that include joint movement and muscles that produce these movements.

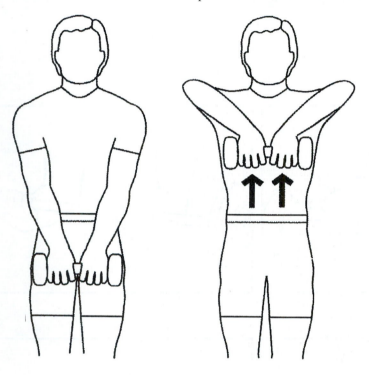

Worksheet no. 2

Analyze this exercise following the procedures explained in this chapter that include joint movement and muscles that produce these movements.

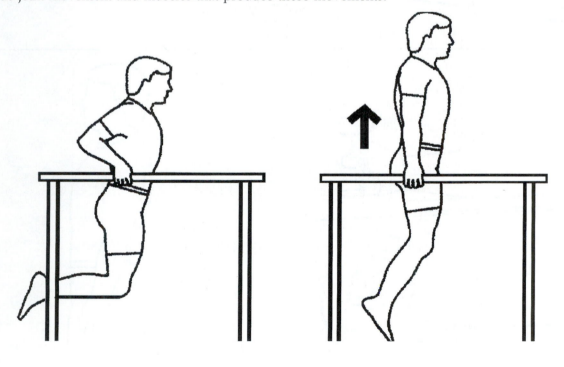

Chapter seven

Worksheet no. 1

Draw and label on the worksheet the anterior hip joint and pelvic girdle muscles. Indicate the origin and insertion of each muscle with an "O" and "I," respectively.

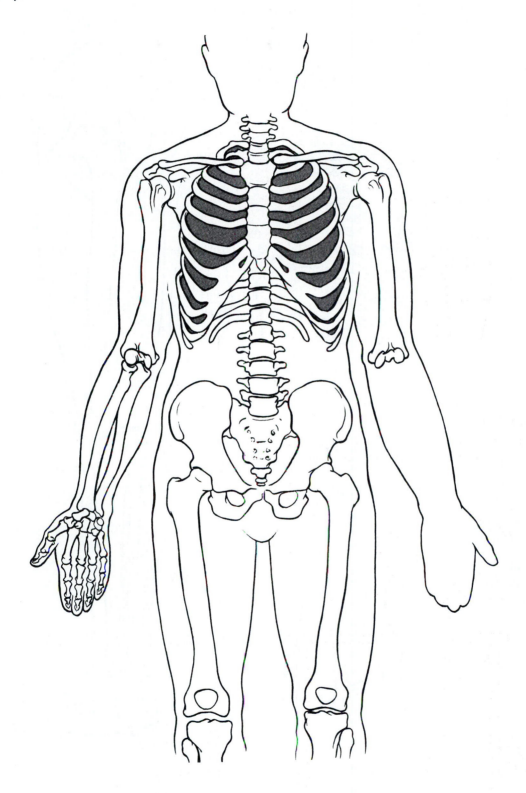

Chapter seven

Worksheet no. 2

Draw and label on the worksheet the posterior hip and pelvic girdle muscles. Indicate the origin and insertion of each muscle with an "O" and "I," respectively.

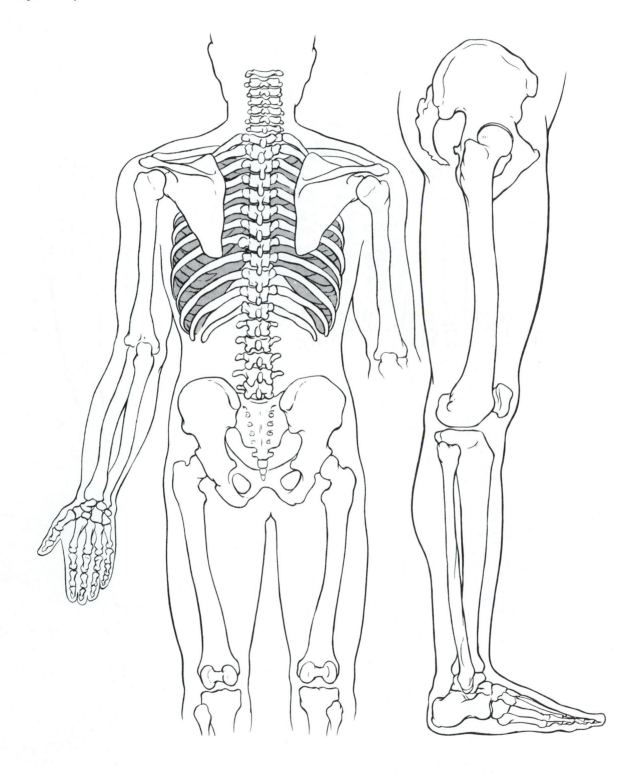

Chapter eight

Worksheet no. 1

Draw and label on the worksheet the knee joint muscles. Indicate the origin and insertion of each muscle with an "O" and "I," respectively.

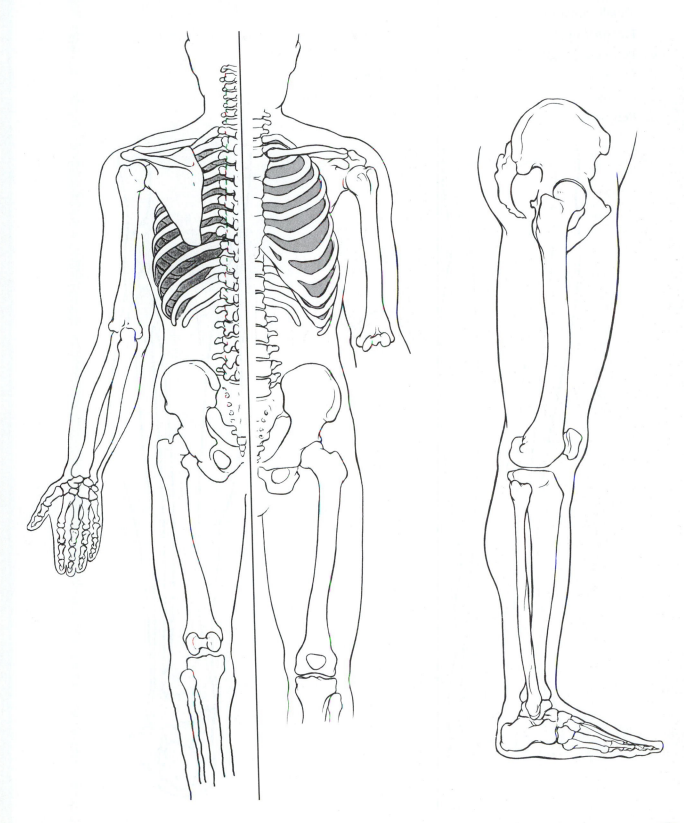

Chapter nine

Worksheet no. 1

Draw and label on the worksheet the following muscles of the ankle and foot. Indicate the origin and insertion of each muscle with an "O" and "I," respectively.

a. Tibialis anterior
b. Extensor digitorum longus
c. Peroneus longus
d. Peroneus brevis
e. Soleus
f. Peroneus tertius
g. Gastrocnemius
h. Extensor hallucis longus
i. Tibialis posterior
j. Flexor digitorum longus
k. Flexor hallucis longus

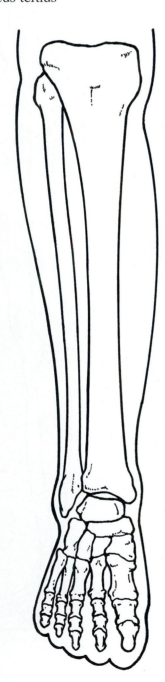
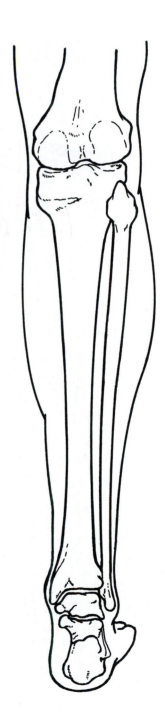

Chapter ten

Worksheet no. 1

Draw and label the following muscles on the skeletal chart. Indicate the origin and insertion of each muscle with an "O" and "I," respectively.

a. Rectus abdominis

b. External oblique abdominal

c. Internal oblique abdominal

d. Sternocleidomastoid

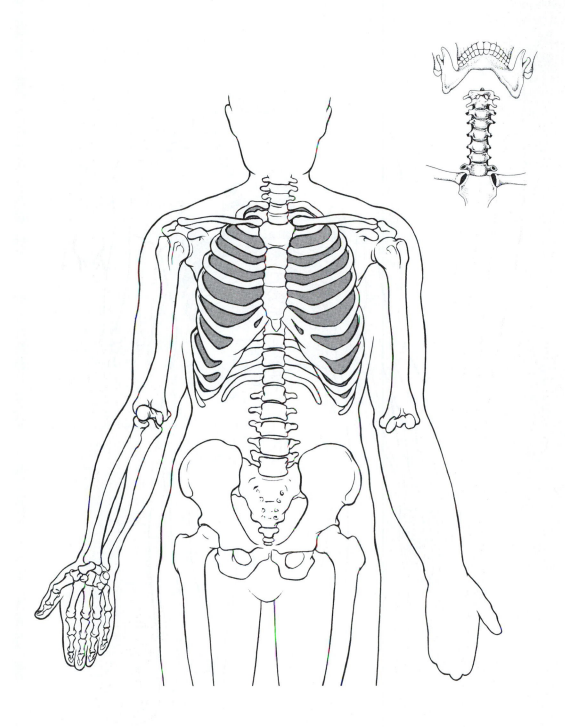

Chapter ten

Worksheet no. 2

Draw and label the following muscles on the skeletal chart. Indicate the origin and insertion of each muscle with an "O" and "I," respectively.

a. Erector spinae

b. Quadratus lumborum

c. Splenius-cervicis and capitis

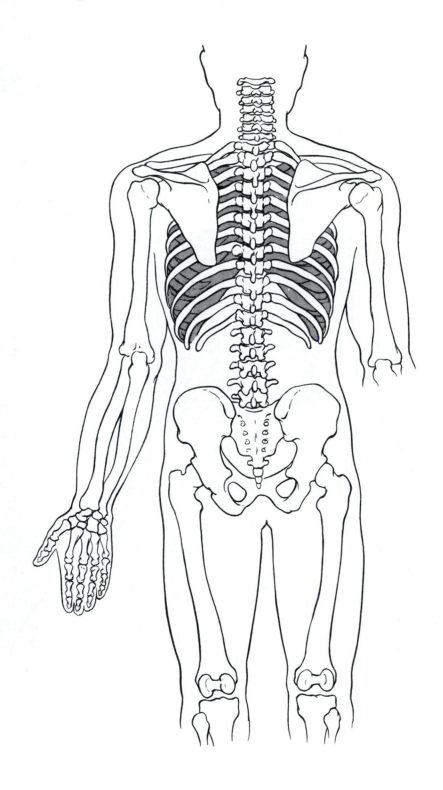

Chapter eleven

Worksheet no. 1

Kinesiology skill analysis table

Phase		Toes	TT/ST	Ankle	Knee	Hip	Lumbar	Cervical	Shou G	G-H	Elbow	Wrist	Fingers
	Position												
	Deg/move												
	Agon/con												
	Position												
	Deg/move												
	Agon/con												
	Position												
	Deg/move												
	Agon/con												
	Position												
	Deg/move												
	Agon/con												
	Position												
	Deg/move												
	Agon/con												

Glossary

abduction   Lateral movement away from the midline of the trunk, as in raising the arms or legs to the side horizontally.

acceleration   The rate of change in velocity.

active insufficiency   Point reached when a muscle becomes shortened to the point that it cannot generate or maintain active tension.

adduction   Movement medially toward the midline of the trunk, as in lowering the arms to the side or legs back to the anatomical position.

agonist   A muscle or muscle group that is described as being primarily responsible for a specific joint movement when contracting.

anatomical position   The position of reference in which the subject is in the standing position, with feet together and palms of hands facing forward.

angle of pull   The angle between the muscle insertion and the bone on which it inserts.

angular displacement   The change in location of a rotating body.

angular motion   Motion involving rotation around an axis.

antagonist   A muscle or muscle group that counteracts or opposes the contraction of another muscle or muscle group.

anteroposterior axis   The axis that has the same directional orientation as the sagittal plane of motion and runs from front to back at a right angle to the frontal plane of motion. Also known as the sagittal or AP axis.

arthrodial joints   Joints in which bones glide on each other in limited movement, as in the bones of the wrist (carpal) or bones of the foot (tarsal).

axis of rotation   The point in a joint which a bone moves or turns about to accomplish joint motion.

balance   The ability to control equilibrium, either static or dynamic.

biarticular muscles   Those muscles whose length from origin to insertion cross two different joints and consequently allow the muscle to perform actions at each joint.

biomechanics   The study of mechanics as it relates to the functional and anatomical analysis of biological systems, especially humans.

center of gravity   The point at which all of the body's mass and weight are equally balanced or equally distributed in all directions.

circumduction   Circular movement of a bone at the joint, as in movement of the hip, shoulder, or trunk around a fixed point.

closed kinetic chain   When the distal end of the extremity is fixed, preventing movement of any one joint unless predictable movements of the other joints in the extremity occur.

concentric contraction   A contraction in which there is a shortening of the muscle.

condyloid joint   Type of joint in which the bones permit movement in two planes without rotation, as in the wrist between the radius and the proximal row of the carpal bones or the second, third, fourth, and fifth metacarpophalangeal joints.

curvilinear motion   Motion along a curved line.

depression   Inferior movement of the shoulder girdle, as in returning to the normal position from a shoulder shrug.

diagonal abduction   Movement by a limb through a diagonal plane away from the midline of the body.

diagonal adduction   Movement by a limb through a diagonal plane toward and across the midline of the body.

diagonal plane   A combination between more than one plane. Less than parallel or perpendicular to the sagittal, frontal, or transverse plane. Also known as oblique plane.

displacement   A change in position or location of an object from its original point of reference.

distal   Farthest from the midline or point of reference; the hand is the most distal part of the upper extremity.

distance   The path of movement, refers to the actual sum length of measurement traveled.

dorsal flexion   Flexion movement of the ankle, resulting in the top of the foot moving toward the anterior tibia bone.

dynamic equilibrium   Occurs when all of the applied and inertial forces acting on the moving body are in balance, resulting in movement with unchanging speed or direction.

dynamics The study of mechanics involving systems in motion with acceleration.

eccentric contraction Contraction in which there is lengthening of a muscle as a result of the force of gravity or a greater force than the contractile force.

eccentric force Force that is applied in a direction not in line with the center of rotation of an object with a fixed axis. In objects without a fixed axis, it is an applied force that is not in line with the object's center of gravity.

elevation Superior movement of the shoulder girdle, as in shrugging the shoulders.

enarthrodial joint Type of joint which permits movement in all planes, as in the shoulder (glenohumeral) and hip joints.

equilibrium State of zero acceleration in which there is no change in the speed or direction of the body.

eversion Turning the sole of the foot outward or laterally, as in standing with the weight on the inner edge of the foot.

extension Straightening movement resulting in an increase of the angle in a joint by moving bones apart, as when the hand moves away from the shoulder.

external rotation Rotary movement around the longitudinal axis of a bone away from the midline of the body. Also known as rotation laterally, outward rotation, and lateral rotation.

fascia Fibrous membrane covering, supporting, connecting, and separating muscles.

first-class lever A lever in which the axis (fulcrum) is between the force and the resistance, as in the extension of the elbow joint.

flexion Movement of the bones toward each other at a joint by decreasing the angle, as at the elbow or knee joint.

follow-through phase Begins immediately after the climax of the movement phase, in order to bring about negative acceleration of the involved limb or body segment; often referred to as the deceleration phase, the velocity of the body segment progressively decreases, usually over a wide range of motion.

force The product of mass times acceleration.

force arm The perpendicular distance between the location of force application and the axis. The shortest distance from the axis of rotation to the line of action of the force. Also known as *the moment arm* or *torque* arm.

frontal plane Bisects the body laterally from side to side, dividing it into front and back halves. Also known as the lateral or coronal plane.

fundamental position Reference position essentially the same as the anatomical position except that the arms are at the sides and facing the body.

ginglymus joint Type of joint which permits a wide range of movement in only one plane such as in the elbow, ankle, and knee joints.

hamstrings A common name given to the group of posterior thigh muscles: biceps femoris, semitendinosus, and semimembranosus.

horizontal abduction Movement of the humerus in the horizontal plane away from the midline of the body.

horizontal adduction Movement of the humerus in the horizontal plane toward the midline of the body.

inertia Resistance to action or change; resistance to acceleration or deceleration. Inertia is the tendency for the current state of motion to be maintained, regardless of whether the body segment is moving at a particular velocity or is motionless.

insertion The point of attachment of a muscle farthest from the midline or center of the body.

internal rotation Rotary movement around the longitudinal axis of a bone toward the midline of the body. Also known as rotation medially, inward rotation, and medial rotation.

intrinsic muscles Muscles that are entirely contained within a specified body part; usually referring to the small, deep muscles found in the foot and hand.

inversion Turning the sole of the foot inward or medially, as in standing with the weight on the outer edge of the foot.

isokinetic Type of dynamic exercise usually using concentric and/or eccentric muscle contractions in which the speed (or velocity) of movement is constant and muscular contraction (usually maximal contraction) occurs throughout the movement.

isometric contraction A type of contraction with little or no shortening of the muscle, resulting in no appreciable change in the joint angle.

isotonic Contraction occurring in which there is either shortening or lengthening in the muscle under tension; also known as a dynamic contraction and may be classified as being either concentric or eccentric.

kinematics The description of motion including consideration of time, displacement, velocity, acceleration, and space factors of a systemës motion.

kinesiology The science of movement, which includes anatomical (structural) and biomechanical (mechanical) aspects of movement.

kinetics The study of forces associated with the motion of a body.

kyphosis Increased anterior concavity of the normal thoracic curve. The lumbar spine may have a reduction of its normal lordotic curve, resulting in a flat-back appearance referred to as lumbar kyphosis.

lateral axis Axis that has the same directional orientation as the frontal plane of motion and runs from side to side at a right angle to the sagittal plane of motion. Also known as the frontal or coronal axis.

lateral epicondylitis A common problem quite frequently associated with gripping and lifting activities commonly known as tennis elbow, which usually involves the extensor digitorum muscle near its origin on the lateral epicondyle.

lateral flexion Movement of the head and or trunk laterally away from the midline; abduction of spine.

lever A rigid bar (bone) that moves about an axis.

ligament   A type of tough connective tissue that attaches bone to bone to provide static stability to joints.

linear displacement   The distance that a system moves in a straight line.

linear motion   Motion along a line; also referred to as translatory motion.

lordosis   Increased posterior concavity of the lumbar and cervical curves.

mass   The amount of matter in a body.

mechanics   The study of physical actions of forces; can be subdivided into statics and dynamics.

medial epicondylitis   An elbow problem associated with the medial wrist flexor and pronator group near their origin on the medial epicondyle; frequently referred to as golfer's elbow.

momentum   The quality of motion, which is equal to mass times velocity.

movement phase   Sometimes known as the acceleration, action, motion, or contact phase, the action part of the skill. Phase in which the summation of force is generated directly to the ball, sport object, or opponent and is usually characterized by near-maximal concentric activity in the involved muscles.

neutralizers   Counteract or neutralize the action of another muscle to prevent undesirable movements; referred to as neutralizing, they contract to resist specific actions of other muscles.

open kinetic chain   When the distal end of the extremity is not fixed to any surface, allowing any one joint in the extremity to move or function separately without necessitating movement of other joints in the extremity.

opposition   Diagonal movement of the thumb across the palmar surface of the hand to make contact with the fingers.

origin   The point of attachment of a muscle closest to the midline or center of the body.

passive insufficiency   Reached when an opposing muscle becomes stretched to the point where it can no longer lengthen and allow movement.

plane of motion   An imaginary two-dimensional surface through which a limb or body segment is moved.

plantar flexion   Extension movement of the ankle, resulting in the foot and/or toes moving away from the body.

preparatory phase   Skill analysis phase, often referred to as the cocking or wind-up phase, used to lengthen the appropriate muscles so that they will be in position to generate more force and momentum as they concentrically contract in the next phase.

pronation   Internally rotating the radius to where it lies diagonally across the ulna, resulting in the palm-down position of the forearm; also used in referring to the combined movements of eversion, abduction, and external rotation of the foot and ankle.

protraction   Forward movement of the shoulder girdle away from the spine; abduction of the scapula.

proximal   Nearest to the midline or point of reference, the first digit of the hand or foot is proximal to the metatarsal.

quadriceps   A common name given to the four muscles of the anterior aspect of the thigh: rectus femoris, vastus medialis, vastus intermedius, and vastus lateralis.

radial flexion   Abduction movement at the wrist of the thumb side of the hand toward the forearm.

reciprocal inhibition   Activation of the motor units of the agonists causes a reciprocal neural inhibition of the motor units of the antagonists, which allows them to subsequently lengthen under less tension. Also referred to as reciprocal innervation.

recovery phase   Skill analysis phase used after follow-through to regain balance and positioning to be ready for the next sport demand.

rectilinear motion   Motion along a straight line.

reduction   Return of the spinal column to the anatomic position from lateral flexion; spine adduction.

resistance arm   The distance between the axis and the point of resistance application.

retraction   Backward movement of the shoulder girdle toward the spine; adduction of the scapula.

rotation   Movement around the axis of a bone, such as the turning inward, outward, downward, or upward of a bone.

rotator cuff   Group of muscles intrinsic to the gleno-humeral joint consisting of the subscapularis, supraspinatus, infraspinatus, and teres minor that is critical in maintaining dynamic stability of the joint.

sagittal plane   Bisects the body from front to back, dividing it into right and left symmetrical halves. Also known as the anteroposterior, or AP, plane.

scoliosis   Lateral curvatures or sideward deviations of the spine.

second-class lever   A lever in which the resistance is between the axis (fulcrum) and the force (effort), as in plantar flexing the foot to raise up on the toes.

sellar joints   Type of reciprocal reception that is found only in the thumb at the carpometacarpal joint and permits ball-and-socket movement, with the exception of rotation.

speed   How fast an object is moving, or the distance an object travels in a specific amount of time.

stability   The resistance to a change in the body's acceleration; the resistance to a disturbance of the body's equilibrium.

stabilizers   Muscles that surround the joint or body part and contract to fixate or stabilize the area to enable another limb or body segment to exert force and move; known as fixators, they are essential in establishing a relatively firm base for the more distal joints to work from when carrying out movements.

stance phase   Skill analysis phase that allows the athlete to assume a comfortable and balanced body position from which to initiate the sport skill; emphasis is on setting the various joint angles in the correct positions with respect to one another and to the sport surface.

static equilibrium The body at complete rest or motionless.

statics The study of mechanics involving the study of systems that are in a constant state of motion, whether at rest with no motion or moving at a constant velocity without acceleration. Involves all forces acting on the body being in balance, resulting in the body being in equilibrium.

supination Externally rotating the radius to where it lies parallel to the ulna, resulting in the palm-up position of the forearm; also used in referring to the combined movements of inversion, adduction, and internal rotation of the foot and ankle.

synchondrosis joint Type of joint separated by a fibrocartilage that allows very slight movement between the bones, such as the symphysis pubis and the costochondral joints of the ribs with the sternum.

syndesmosis joint Type of joint held together by strong ligamentous structures that allow minimal movement between the bones, such as the coracoclavicular joint and the inferior tibiofibular joint.

synergist Muscles that assist in the action of the agonists but are not primarily responsible for the action; known as guiding muscles, they assist in refined movement and rule out undesired motions.

tendon Fibrous connective tissue, often cordlike in appearance, that connects muscles to bones and other structures.

third-class lever A lever in which the force (effort) is between the axis (fulcrum) and the resistance, as in flexion of the elbow joint.

torque Moment of force. The turning effect of an eccentric force.

transverse plane Divides the body horizontally into superior and inferior halves; also known as horizontal plane.

trochoidal joint Type of joint with a rotational movement around a long axis, as in rotation of the radius at the radioulnar joint.

ulnar flexion Adduction movement at the wrist of the little finger side of the hand toward the forearm.

velocity Includes the direction and describes the rate of displacement.

vertical axis Runs straight down through the top of the head and is at a right angle to the transverse plane of motion; also known as the longitudinal or long axis.

Illustration credits

CHAPTER 1

1.1, Van de Graff KM: *Human anatomy*, ed 4, 1995, McGraw-Hill Companies, Inc., New York.; **1.2**, **1.3**, Thibodeau GA: *Anatomy and physiology*, St. Louis, 1987, Mosby; **1.4**, Arnheim DD, Prentice WE: *Principles of athletic training*, ed 10, 2000 Mc-Graw-Hill Companies, Inc., New York.; **1.5 through 1.8, 1.10**, Booher JM, Thibodeau GA: *Athletic injury assessment*, ed 4, 2000, McGraw-Hill Companies, Inc., New York; **1.13**, Booher JM, Thibodeau GA: *Athletic injury assessment*, ed 3, St. Louis, 1994, Mosby; **1.9**, Seeley R, et al: *Anatomy and physiology*, ed 3, St. Louis, 1995, Mosby; **1.11**, Arnheim DD, Prentice WE: *Principles of athletic training*, ed 9; St. Louis, 1997, McGraw-Hill Companies, Inc., New York., **1.12**, R. T. Floyd; **1.14**, John Hood.

CHAPTER 2

2.1, 2.2, 2.9, Linda Kimbrough; **2.4**, Hall SJ: *Basic biomechanics*, ed 3, 1999, McGraw-Hill Companies, Inc., New York.; **2.3**, John Hood; **2.4 through 2.8**, Ernest W. Beck.

CHAPTER 3

3.1, 3.7, 3.9, Linda Kimbrough; **3.2, 3.3, 3.5**, John Hood; **3.4**, Booher JM, Thibodeau GA: *Athletic injury assessment*, ed 2, St. Louis, 1989, Mosby; **3.6, 3.8, 3.14 through 3.16**, Ernest W. Beck; **3.10 through 3.13**, Ernest W. Beck with inserts by Linda Kimbrough.

CHAPTER 4

4.1, 4.2 A, 4.6 through 4.13, Linda Kimbrough; **4.2 B**, Van de Graff KM: *Human anatomy*, ed 4, 1995, McGraw-Hill Companies, Inc., New York.; **4-3**, John Hood; **4.4, 4.5**, Thibodeau GA: *Anatomy and physiology*, St. Louis, 1987, Mosby.

CHAPTER 5

5.1, Anthony CP, Kolthoff NJ: *Textbook of anatomy and physiology*, ed 9, St. Louis, 1975, Mosby; **5.2, 5.20**, Van de Graff KM: *Human anatomy*, ed 4, 1995, McGraw-Hill Companies, Inc., New York.; **5.3, 5.5 through 5.19**, Linda Kimbrough; **5.4**, John Hood.

CHAPTER 6

6.1, R. T. Floyd; **6.2 through 6.4, 6.6**, Lisa Floyd; **6.5**, John Hood; **6.7 A–C, 6.8 A–B, 6.9 A–B**, Ron Carlberg.

CHAPTER 7

7.1, 7.3, 7.6 through 7.22, Linda Kimbrough; **7.2, 7.24**, Anthony CP, Kolthoff NJ: *Textbook of anatomy and physiology*, ed 9, St. Louis, 1975, Mosby; **7.4 A–F, 7.5 A–D**, John Hood; **7.23**, Ernest W. Beck.

CHAPTER 8

8.1, Anthony CP, Kolthoff NJ: *Textbook of anatomy and physiology*, ed 9, St. Louis, 1975, Mosby; **8.2 A–D**, John Hood; **8.3 through 8.8**, Linda Kimbrough.

CHAPTER 9

9.1, 9.2, 9.4, Anthony CP, Kolthoff NJ: *Textbook of anatomy and physiology*, ed 9, St. Louis, 1975, Mosby; **9.3, 9.18**, Van de Graff KM: *Human anatomy*, ed 4, 1995, McGraw-Hill Companies, Inc., New York.; **9.5**, Seeley R, et al: *Anatomy and physiology*, ed 3, St. Louis, 1995, Mosby; **9.6 A–F**, John Hood; **9.7 through 9.13, 9.15 through 9.17**, Ernest W. Beck; **9.14**, Linda Kimbrough.

CHAPTER 10

10.1, 10.11, Seeley R, et al: *Anatomy and physiology*, ed 3, St. Louis, 1995, Mosby; **10.2 E–F**, Anthony CP, Kolthoff NJ: *Textbook of anatomy and physiology*, ed 9, St. Louis, 1975, Mosby; **10.2 A–D, 10.10, 10.13 through 10.16, 10.18 through 10.21**, Linda Kimbrough; **10.3**, Hole J: *Human anatomy and physiology*, ed 6, 1993, McGraw-Hill Companies, Inc., New York., **10.4, 10.7**, Lindsay D: *Functional anatomy*, ed 1, St. Louis, 1996, Mosby, **10.5**, Thibodeau GA, Patton KT: *Anatomy and physiology*, ed 9, St. Louis, 1993, Mosby; **10.6 A–H**, John Hood; **10.8, 10.12**, Van de Graff KM: *Human anatomy*, ed 4, 1995, McGraw-Hill Companies, Inc., New York.; **10.9, 10.17**, Ernest W. Beck.

CHAPTER 11

11.1, 11.3 through 11.6, Lisa Floyd; **11.2**, John Hood; **11.7**, Efi Medical Systems, San Diego, CA; **11.8, 11.9 A–B, 11.10 A–B**, Ron Carlberg.

CHAPTER 12

12.1, 12.6, Hall SJ: *Basic biomechanics*, ed 3, 1999, McGraw-Hill Companies, Inc., New York.; **12.2, 12.3, 12.4**, Booher JM, Thibodeau GA: *Athletic injury assessment*, ed 2, St. Louis, 1989, Mosby, Hall SJ: *Basic biomechanics*, ed 3, 1999, McGraw-Hill Companies, Inc., New York..

Index

Page numbers in *italics* refer to illustrations.